# ADVANCED COMPOSITES
## for AEROSPACE, MARINE, and LAND APPLICATIONS

**Cover Photograph:** Image of the sintered hybrid preform having 20 vol. % of fiber hybridized with 20% vol. of SiCp. Further details can be found in "Fabrication and Characterization of a Hybrid Functionally Graded Metal-Matrix Composite using the Technique of Squeeze Infiltration," by K. M. Sree Manu, V.G. Resmi, Prince Joseph, T.P.D. Rajan, B.C. Pai, and T.S. Srivatsan.

# TMS 2014
## 143rd Annual Meeting & Exhibition

New proceedings volumes from the TMS2014 Annual Meeting, available from publisher John Wiley & Sons:

- 5th International Symposium on High-Temperature Metallurgical Processing
- Advanced Composites for Aerospace, Marine, and Land Applications
- Celebrating the Megascale: Proceedings of the Extraction and Processing Division Symposium on Pyrometallurgy in Honor of David G.C. Robertson
- Characterization of Minerals, Metals, and Materials 2014
- Energy Technology 2014: Carbon Dioxide Management and Other Technologies
- EPD Congress 2014
- Light Metals 2014
- Magnesium Technology 2014
- Rare Metal Technology 2014
- Shape Casting: 5th International Symposium 2014
- TMS 2014 Supplemental Proceedings

To purchase any of these books, visit **www.wiley.com**.

TMS members: Log in to the Members Only area of **www.tms.org** and learn how to get your discount on these and other books offered by Wiley.

# ADVANCED COMPOSITES
## for AEROSPACE, MARINE, and LAND APPLICATIONS

*Proceedings of a symposium sponsored by*
The Minerals, Metals & Materials Society (TMS)

*held during*

**TMS2014**
143rd Annual Meeting & Exhibition

February 16-20, 2014
San Diego Convention Center
San Diego, California, USA

*Edited by:*
Tomoko Sano • T. S. Srivatsan
Michael W. Peretti

WILEY | TMS

Copyright © 2014 by The Minerals, Metals & Materials Society.
All rights reserved.

Published by John Wiley & Sons, Inc., Hoboken, New Jersey.
Published simultaneously in Canada.

No part of this publication may be reproduced, stored in a retrieval system, or transmitted in any form or by any means, electronic, mechanical, photocopying, recording, scanning, or otherwise, except as permitted under Section 107 or 108 of the 1976 United States Copyright Act, without either the prior written permission of The Minerals, Metals, & Materials Society, or authorization through payment of the appropriate per-copy fee to the Copyright Clearance Center, Inc., 222 Rosewood Drive, Danvers, MA 01923, (978) 750-8400, fax (978) 750-4470, or on the web at www.copyright.com. Requests to the Publisher for permission should be addressed to the Permissions Department, John Wiley & Sons, Inc., 111 River Street, Hoboken, NJ 07030, (201) 748-6011, fax (201) 748-6008, or online at http://www.wiley.com/go/permission.

Limit of Liability/Disclaimer of Warranty: While the publisher and author have used their best efforts in preparing this book, they make no representations or warranties with respect to the accuracy or completeness of the contents of this book and specifically disclaim any implied warranties of merchantability or fitness for a particular purpose. No warranty may be created or extended by sales representatives or written sales materials. The advice and strategies contained herein may not be suitable for your situation. You should consult with a professional where appropriate. Neither the publisher nor author shall be liable for any loss of profit or any other commercial damages, including but not limited to special, incidental, consequential, or other damages.

Wiley also publishes books in a variety of electronic formats. Some content that appears in print may not be available in electronic formats. For more information about Wiley products, visit the web site at www.wiley.com. For general information on other Wiley products and services or for technical support, please contact the Wiley Customer Care Department within the United States at (800) 762-2974, outside the United States at (317) 572-3993 or fax (317) 572-4002.

Library of Congress Cataloging-in-Publication Data is available.

ISBN 978-1-118-88891-9

Printed in the United States of America.

10 9 8 7 6 5 4 3 2 1

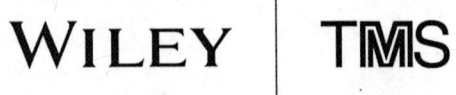

# TABLE OF CONTENTS
## Advanced Composites for Aerospace, Marine, and Land Applications

Preface ............................................................................................... ix
About the Organizers ........................................................................ xi
Session Chairs ................................................................................... xv

## Processing and Design of Composites

Deformation Behaviour of Aluminium Alloy AA6061-10% Fly Ash
Composites for Aerospace Application ............................................. 3
   *A. Bhandakkar, R. Prasad, and S. Sastry*

Effect of Composition of $B_4C$-Aluminum Composites on Mechanical
Properties and Resistance Corrosion ............................................... 23
   *L. Vázquez, E. Hernández, A. Altamirano, V. Cortés, E. Garfias,
   E. Refugio and M. Vite*

Bacterial Cellulose Enhances Beta Phase in PVDF ........................ 35
   *V. Verma, P. Rajesh, S. Bodkhe, and S. Kamle*

Geopolymer from Industrial Wastes: A Construction Material for
$22^{nd}$ Century ..................................................................................... 43
   *P. Rana, R. Dash, and R. Ganguly*

Synthesis of Composite $TaC$-$TaB_2$ Powders .................................. 55
   *B. Mehdikhani, G. Borhani, S. Bakhshi, and H. Baharvandi*

## Characterization of Composite Microstructures and Phases

Laser Deposited In-Situ TiC Reinforced Nickel Matrix Composites:
Microstructure and Tribological Properties .................................... 67
   *T. Borkar, S. Gopagoni, R. Banerjee, J. Hwang, and J. Tiley*

Use of Squeeze Infiltration Processing for Fabricating Micro Silica
Reinforced Aluminum Alloy-Based Metal Matrix Composite ........................... 81
   *K. Manu, V. Resmi, T. Rajan, P. Joseph, B. Pai, and T. Srivatsan*

Fabrication and Characterization of a Hybrid Functionally Graded
Metal-Matrix Composite Using the Technique of Squeeze Infiltration ............. 91
   *K. Manu, V. Resmi, P. Joseph, T. Rajan, B. Pai, and T. Srivatsan*

The Microstructure and Mechanical Properties of Magnetic Shape Memory
Alloys $NiCo_{40+x}Al_{30-x}[X=0、3、6、10]$ ........................................................ 101
   *J. Ju, F. Xue, J. Zhou, J. Bai, and H. Liu*

Metal Matrix Composites Directionally Solidified ........................................... 115
   *A. Ares and C. Schvezov*

## Mechanical and Material Property Evaluation

A New Class of Metal Nanocomposites with Superior Mechanical
Properties: Unusual Thermal Expansion in NbTi-Nanowires/NiTi-Matrix
Composite ......................................................................................................... 127
   *S. Hao, L. Cui, S. Wang, D. Jiang, C. Yu, D. Brown, and Y. Ren*

Data-Fusion NDE for Progressive Damage Quantification in Composites ...... 137
   *J. Cuadra, P. Vanniamparambil, K. Hazeli, I. Bartoli,*
   *and A. Kontsos*

Progressive Failure Analysis of Polymer Composites Using a Synergistic
Damage Mechanics Methodology ................................................................... 147
   *J. Montesano and C. Singh*

Computational Prediction of Mechanical Properties of Glassy Polymer
Blends and Thermosets .................................................................................... 157
   *D. Rigby, P. Saxe, C. Freeman, and B. Leblanc*

Multi Scale Characterization of SiC/SiC Composite Materials ....................... 173
   *D. Frazer, M. Abad, C. Back, C. Deck, and P. Hosemann*

Processing Fracture Toughness and Damage Mechanics Studies of MMC
for Aerospace ................................................................................................... 185
   *A. Bhandakkar, R. Prasad, and S. Sastry*

# Interface and Bonding of Composite Systems

Multi-Scale Modeling of Continuous Ceramic Fiber Reinforced Aluminum
Matrix Composites........................................................................................203
   B. McWilliams and C. Yen

Development of Percussion Diagnostics in Evaluating 'Kiss' Bonds between
Composite Laminates ..................................................................................213
   S. Poveromo and J. Earthman

Enhanced Mechanical Performance of Woven Composite Laminates Using
Plasma Treated Polymeric Fabrics................................................................231
   T. Walter, A. Bujanda, V. Rodriguez-Santiago, J. Yim, J. Baeza,
   and D. Pappas

Forming Limit Diagram of Steel/Polymer/Steel Sandwich Systems for the
Automotive Industry.....................................................................................243
   M. Harhash and H. Palkowski

The Wettability of $TiC_X$ by Titanium-Aluminum Alloys at 1758 K.................255
   X. Liu, X. Lv, C. Bai, and C. Li

Author Index................................................................................................265

Subject Index...............................................................................................267

# PREFACE

This volume contains the papers presented at the international symposium titled "Advanced Composites for Aerospace, Marine, and Land Applications" held during the TMS 2014 Annual Meeting & Exhibition in San Diego, California, USA, February 16-20, 2014. The four-session symposium was sponsored by the Composite Materials Committee of TMS (The Minerals, Metals & Materials Society).

This symposium is the first in a planned and forthcoming series on the topic of advanced composites and was reasonably well-represented with abstracts from engineers, technologists, and scientists spanning the domains of academia, research laboratories, and industries located both within the United States and a few countries overseas. Over 30 abstracts were approved for presentation at the symposium and were divided into four sessions:

**Session 1:** Processing and Design of Composites
**Session 2:** Characteristics of Composite Microstructures and Phases
**Session 3:** Mechanical and Material Property Evaluation
**Session 4:** Interface and Bonding of Composite Systems

The abstracts that were chosen for presentation cover a broad spectrum of topics that represent the truly diverse nature of the field of composite materials. In recent years, composite materials have grown in strength, stature, and significance to become a key material of enhanced scientific interest, and resultant research into understanding their behavior for selection and safe use in a wide spectrum of technology-related applications. In this symposium, research describing the latest advances in composite materials specifically for aerospace, maritime, or land applications were emphasized and presented. We have made every attempt to bring together individuals who could put forth recent advances in their research while concurrently enhancing our prevailing understanding of aspects related to the science, engineering and far-reaching technological applications of composite materials. We, the symposium organizers, extend our warmest thanks and appreciation to both the authors and session chairmen for their enthusiastic commitment and participation.

We also extend our most sincere thanks and appreciation to the elected and governing representatives of TMS for their understanding and acknowledgement of our interest, and timely approval of our request to organize this intellectually stimulating event. An overdose of thanks, gratitude, and appreciation are extended to Ms. Trudi Dunlap (Programming Manager at The Minerals, Metals & Materials Society (TMS) for her sustained attention, assistance, interest, involvement, and timely participation stemming from understanding. This certainly enabled in ensuring a timely execution of the numerous intricacies related to both orchestration and layout of this symposium from the moment following its approval and up until compilation and publication of the bound volume (proceedings). At moments of need, we the symposium organizers did find her presence, at the TMS headquarters (Warrendale, PA), and participation to be a pillar of support, courteous, understanding, professional and almost always enthusiastically receptive and helpful. Special thanks and much deserved appreciation is extended to Ms. Patricia Warren (Programming and Proceedings Specialist at (TMS) for her patience, understanding, and much valued and desired attention in compilation and presentation of this bound volume. Also, the timely publication of this bound volume would not have been possible without the cooperation of the authors and the publishing staff headed by the dynamic, diligent, and dedicated Mr. Bob Demmler at TMS (Warrendale, PA, USA).

We truly hope that this bound volume will provide engineers, scientists, and technologists with both new perspectives and directions in their research endeavors with the purpose of evaluating and understanding the behavior of composite materials so as to enable their selection and use for varied applications in the sectors spanning aerospace, marine, and land-based components and products.

**Tomoko Sano**
U.S. Army Research Laboratory
Weapons and Materials Research Directorate
Materials Response and Design Branch
Aberdeen Proving Ground, MD 21005

**T. S. Srivatsan**
Division of Materials Science and Engineering
Department of Mechanical Engineering
The University of Akron
Akron, OH 44325

**Michael W. Peretti**
Director, Advanced Programs
GE Aviation
1 Neumann Way
Cincinnati, OH 45215

# SYMPOSIUM ORGANIZERS

**Tomoko Sano**, Materials Engineer, U.S. Army Research Laboratory, received her M.S. in Materials Science and Engineering at Carnegie Mellon University in 2001 and her Ph.D. in Materials Science and Engineering at Carnegie Mellon University in 2005. After a postdoctoral fellowship at the U.S. Army Research Laboratory, she was appointed the Technical Assistant to the Director of the Weapons and Materials Research Directorate, where she was awarded three special act awards for the technical execution of directorate level assignments and programmatic actions. She is currently a materials engineer in the Materials and Manufacturing Science Division of the Weapons and Materials Research Directorate of the U.S. Army Research Laboratory, Aberdeen Proving Ground, Maryland.

Dr. Sano has been the first author of over 20 technical publications, and presented over 20 conference presentations. She is an active member of the TMS Composite Materials Committee, and has been the JOM liaison in 2012 and 2014.

**T.S. Srivatsan**, Professor of Materials Science and Engineering in the Department of Mechanical Engineering at the University of Akron. He received his graduate degrees (Master of Science in Aerospace Engineering in 1981 and Doctor of Philosophy in Mechanical Engineering in 1984) from Georgia Institute of Technology. Dr. Srivatsan joined the faculty in the Department of Mechanical Engineering at the University of Akron in August 1987. Since joining, he has instructed undergraduate and graduate courses in the areas of Advanced Materials and Manufacturing Processes, Mechanical Behavior of Materials, Fatigue of Engineering Materials and Structures,  Fracture Mechanics, Introduction to Materials Science and Engineering, Mechanical Measurements, Design of Mechanical Systems and Mechanical Engineering Laboratory. His research areas currently span the fatigue and fracture behavior of advanced materials to include monolithic(s), intermetallic, nano-materials and metal-matrix composites; processing techniques for advanced materials and nanostructure materials; interrelationship between processing and mechanical behavior; electron microscopy; failure analysis; and mechanical design. His funding comes primarily from both industries and government and is of the order of a few millions of dollars since 1987. A synergism of his efforts has helped in many ways to advancing the science, engineering and technological applications of materials.

Dr. Srivatsan has authored/edited/co-edited 51 books in areas cross-pollinating mechanical design; processing and fabrication of advanced materials; deformation, fatigue and fracture of ordered intermetallic materials; machining of composites; failure analysis; and technology of rapid solidification processing of materials. He serves as co-editor of *International Journal on Materials and Manufacturing Processes* and is on the editorial advisory board of journals in the domain of Materials Science and Engineering. His research has enabled him to deliver over 200 technical presentations in national and international meetings and symposia, technical/professional societies, and research and educational institutions. He has authored and co-authored over 600 archival publications in international journals, chapters in books, proceedings of national and international conferences, reviews of books, and technical reports. In recognition of his contributions and their impact on furthering science, technology, and education he has been elected Fellow of the American Society for Materials, International (ASM Int.); Fellow of American Society of Mechanical Engineers (ASME); and Fellow of American Association for the Advancement of Science (AAAS). He has also been recognized as Outstanding Young Alumnus of Georgia Institute of Technology, and outstanding Research Faculty, the College of Engineering at the University of Akron. He offers his knowledge in research services to the U.S. government (U.S. Air Force and U.S. Navy), national research laboratories, and industries related to aerospace, automotive, power-generation, leisure-related products, and applied medical sciences. He has the distinct honor of being chosenfor inclusion in *Who's Who in American Education, Who's Who in the Midwest, Who's Who in Technology, Who's Who in the World, Who's Who in America;, Who's Who in Science and Engineering, Who's Who among America's Teachers, and Who's Who among Executives and Professionals* (Cambridge).

**Michael W. Peretti**, Director, Advanced Programs at GE Aviation, Cincinnati, Ohio. He received his Bachelor of Engineering Degree in Metallurgical Engineering from Stevens Institute of Technology in 1982, and his Master of Science Degree in Material Science from the University of Pittsburgh in 1999. His 30-plus years of experience with aerospace propulsion materials includes processing of conventional and powder metallurgy iron, nickel, and cobalt alloys, failure analysis, near-net-shape manufacturing, and applications and technology development for polymer and ceramic composite materials.

He has authored more than 10 technical publications in the fields of gas and rotary atomization, nickel and titanium alloy development and processing, and specialty alloys, and numerous commentary articles on Titanium Processing. He holds four U.S. patents. He has served TMS on the Titanium and Composites Committees in the Structural Materials Division, also serving as JOM liaison and guest editor for the Titanium Committee.

# SESSION CHAIRS

**Processing and Design of Composites**
Liang Dong
Michael W. Peretti

**Characterization of Composite Microstructures and Phases**
Tushar Borkar
T.S. Srivatsan

**Mechanical and Material Property Evaluation**
Yang Ren
C.K.H. Dharan

**Interface and Bonding of Composite Systems**
Tomoko Sano
Brandon McWilliams

# PROCESSING AND DESIGN OF COMPOSITES

# DEFORMATION BEHAVIOUR OF ALUMINIUM ALLOY AA6061-10% FLY ASH COMPOSITES FOR AEROSPACE APPLICATION

Ajit Bhandakkar[1], R C Prasad[1] and Shankar M L Sastry[2]

[1]Department of Metallurgical Engineering and Materials Science
**Indian Institute of Technology** (IIT) Bombay
Bombay, **India**

[2]Mechanical, Aerospace and Structural Engineering
**Washington University** at St. Louis
St. Louis, **USA**

## Abstract

Aluminum fly ash metal matrix composites (MMCs) find important applications in aerospace where specific stiffness is important. The aluminum fly ash metal matrix composites (ALFA) can be processed through low cost processing route ie by liquid metallurgy - stir casting. These ALFA composites have limited ductility which makes secondary mechanical processing very difficult, the strain rate and temperature for deformation processing is generally set by trial and error methods. The development of processing maps for the optimization of processing parameters such as strain rate and temperature has been found to be very useful, The aim of the present investigation is to study the constitutive flow behavior of AA6061 aluminum alloy-10 vol. % fly ash particulate MMC under hot-working conditions and to generate a processing map for optimization of hot workability.

The mechanical response of AA6061aluminum fly ash metal matrix composite was investigated by means of hot compression tests. The flow stress curves were obtained in the temperature and strain rate ranges of 200–500 °C and 0.5–0.01 $s^{-1}$, respectively in order to obtain the processing and stability maps. The different domains in the processing maps may be correlated with specific micro structural mechanisms and the "safe" mechanisms were chosen for processing. The hot deformation behavior of the matrix material AA6061 was also evaluated for the purpose of comparison with that of the aluminum fly ash metal matrix composites. All the zones of flow instability, micro structural and failure mechanisms involved in the hot working of the MMC were studied by optical microscope and scanning electron microscope.

**Keywords**: Processing Maps, Aluminium alloy AA6061,Fly Ash Composites

## Introduction

Aluminum alloy-based metal matrix composites (MMCs) find important applications where specific stiffness is important and are produced by stir casting as well as powder metallurgy (PM) routes. As these MMCs have limited ductility, mechanical processing is a critical step and strain rate for deformation processing is generally done by trial and error methods, although recently, the use of processing maps for the optimization of processing parameters has been found to be very useful, Tuler and Klimowicz [2] developed processing maps for aluminum alloy cast MMCs and conducted successful forging trials validating the maps. The aim of the present investigation is to study the constitutive flow behavior of 6061 aluminum alloy-10 vol. percent fly ash particulate MMC under hot-working conditions and to generate a processing map for optimization of hot workability.

The map is interpreted on the basis of the Dynamic Materials Model (DMM), which is reviewed by Gegel *et* al. [3] In this model, the work piece subjected to hot working is considered as a nonlinear dissipator of power and the instantaneous power dissipated at a given strain rate may be considered to consist of two parts: G content, representing the temperature rise, and J co-content, representing the dissipation through metallurgical processes.

The factor that partitions power between J and G is the strain-rate sensitivity (m) of flow stress, and the J co-content is given by $J = 2m/(m + 1)$ [1] For an ideal linear dissipater, $J = Jmax$. The efficiency of power dissipation of a nonlinear dissipater may be expressed as a dimensionless parameter $r = J/Jmax$. The variation of r with temperature and strain rate represents the characteristics of power dissipation through micro structural changes in the work piece and constitutes a processing map. The different domains in the map may be correlated with specific micro structural mechanisms, and the "safe" mechanisms may be chosen for processing. In this study, the hot deformation behavior of the matrix material AA6061 is also evaluated for the purpose of comparison with that of the MMC. This would help in understanding the micro structural mechanisms involved in the hot working of the MMC.

Monolithic materials often find limitations on increasing performance requirements in various environments. Hence there has been need to find new materials with tailor made properties. Aluminum alloy-based metal matrix composites (MMCs) find important applications where specific stiffness is important and are produced by stir casting as well as powder metallurgy (PM) routes. As these MMCs have limited ductility, mechanical processing is a critical step and strain rate for deformation processing is generally done by trial and error methods, although recently, the use of processing maps for the optimization of processing parameters has been found to be very useful.

Fly ash is a waste of coal fired power plants which is cheap and abundantly available .When incorporated in metal matrix it offers value addition. However there are issues related to processing and product performance of composites. The qualification for industrial applications is based on understanding host of properties, environmental effects on them and understanding the micro mechanisms of fracture. The present investigation addresses the issues related to processing and properties of fly ash incorporated MMCs.

## EXPERIMENTAL

### Processing of Composites Using Stir Casting Route

Castings of AA6061 base alloy and their composites with 5 wt.% and 10 wt.% fly-ash have been made through the liquid metal stir casting route. The composites were processed using as received P-60 fly ash of 80 micron mean particle size. The aluminium AA6061 alloy was melted at 800°C in the electrical resistance furnace having temperature controller. The casting was obtained in the form of billet. The volume fraction of fly-ash in the composites was varied during stirring of molten metal. Fly ash was heated to about 600 °C for about 2 h before mixing into the melt. The mixing and the casting operation were carried out at a temperature above the liquidus temperature of the AA6061 alloy. The temperatures at which the mixing and casting operations were done were varied to ascertain its effect on the uniform distribution of the reinforcement and formation of agglomerates. Fly ash addition was carried out at a rate of 10 g / min in all the cases and the composite melt was mixed for about 30 minutes. The melt was stirred using a graphite stirrer driven by an electrical motor at 200 rpm during the mixing of the reinforcement. After completion of the mixing procedure, the composite melt was stirred for 5 minutes before it was poured into the moulds. The picture depicting the stirring of the melt is shown in Fig. 1

**Figure 1.** Experimental setup for composite manufacturing through liquid stir casting

The composites were cast in two types of moulds, a rectangular mould and a cylindrical mould. Before Casting, the mould surface was heated sufficiently to remove any traces of moisture. The photographs of the castings are shown in Fig.2.

**Figure 2.** Photographs of the molds and castings of MMCs

## Secondary processing

The as-cast composite billets were extruded for (64 % reduction) at 450 °C (Soaking for 4 hrs) in order to get rid of the porosities induced during primary processing. It also improves the distribution of the reinforcement in the matrix. Secondary processing improves distribution of fly ash reinforcement in the matrix, imparts directional properties, whereby mechanical properties are improved. The hot extrusion details of Metal Matrix Composite (AA6061 + fly ash ) are shown in Table 4.

Table 1. Extrusion ratio used for secondary processing of AA6061/fly ash composites

| Material | Initial Diameter(mm) | Final Diameter | % Reduction |
|---|---|---|---|
| AA 6061 base alloy | 49.5 | 17.74 | 64.16 |
| AA6061 +5%P60 | 49.5 | 17.74 | 64.16 |
| AA6061 + 10%P60 | 49.5 | 17.74 | 64.16 |

The aluminum metal matrix composites were ECAP at 500 °C in order to disperse the fly ash reinforcement uniformly in the aluminum matrix.

Figure 3.   Processing of composites by ECAP

## Deformation mechanism maps

A processing map is an explicit representation of the response of a material in terms of micro structural mechanisms, to the imposed process parameters. These are developed on the basis of Dynamic materials model. The input to generate process maps is the experimental data of flow stress as a function of temperature strain rate and strain. In a compression test on a cylindrical specimen it is easy to obtain a constant true strain rate using experimental decay of the actuator speed .It is easy to measure adiabatic temperature rise directly on the specimen and conduct the tests under isothermal conditions.

Dynamic materials modeling, first developed by Prasad and Sasidhara, is based on fundamental properties of continuum mechanics of large plastic flow, physical system modeling, and irreversible thermodynamics. In this model, the work piece during hot deformation is considered to be a dissipater of power. The dissipated power is converted into thermal and micro structural forms. The factor that partitions power in to these two forms is the strain rate sensitivity '$m$' of the flow stress which can be evaluated as follows:

$$\frac{\Delta \log \bar{\sigma}}{\Delta \log \dot{\varepsilon}} = m$$

Where $\sigma$ is the flow stress and $\dot{\varepsilon}$ is the strain rate.

The variation of the parameter m with temperature and strain rate constitutes the instability map in which, regions of negative parameter represent the flow instabilities. The super position of the power dissipation map over instability map gives a processing map, which may be used to characterize different domains where different micro structural mechanisms occur as well as the regimes that exhibit flow instability.

## Test procedure for development of processing maps

AA 6061 -10%FA composite aluminum alloy was fabricated by liquid metallurgy stir casting route. The billets were vacuum hot pressed and cylindrical compression specimens of 10-mm diameter and 15-mm height were machined such that their faces were parallel. Concentric grooves of about 0.5 mm depth were engraved on the specimen faces to facilitate the retention of the lubricant. A 1mm 45 deg chamfer was given to the edges of the faces to avoid fold over in the initial stages of the compression. A 0.5mm-diameter was drilled to a depth of 5 mm at half the height of each specimen for the insertion of a thermocouple. Hot compression tests were conducted in the temperature range of 250 to 500 °C and in the strain-rate range of 0.001 to 1 s for the AA6061 fly ash composites and for the matrix material.

A computer-controlled servo hydraulic machine (MTS, United States) was used in this investigation. The deformed specimens were sectioned vertically, and the surface was prepared for metallographic examination using standard metallographic techniques. The polished specimens were etched with a solution consisting of 2 ml $HBF_4$, 3 ml HCl, 5 ml $HNO_3$, and 190 ml $H_2O$.

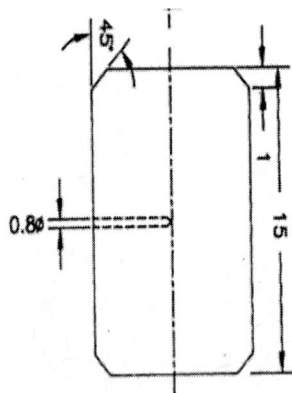

**Figure 4 Geometry of Compression Specimen**

The load vs. stroke data are then converted to true stress- true plastic strain curves using the standard equations given below:

$$\sigma = P/A_o (1 - e)$$

$$\varepsilon = - \ln (h_o/h)$$

Here $\sigma$ is true stress, P is load, $A_o$ is original area of crossection of the sample, e is engineering strain and $\varepsilon$ is true strain.
Then the flow stress data obtained is corrected for adiabatic temperature rise by applying linear interpolation between $\log(\sigma)$ and (1/T) values. Cubic spline interpolation is carried out to compute the flow stress values at finer temperature and strain rate intervals, using experimental data points as knots. The strain rate sensitivity (m) is calculated using the following equation:

$$\frac{\Delta \log \overline{\sigma}}{\Delta \log \dot{\varepsilon}} = m$$

Consequently, efficiency of power dissipation ($\eta$) and instability parameter ($\xi(\dot{\varepsilon})$) are calculated from the given relations and plotted.

$$\eta = \frac{2m}{m+1}$$

$$\xi(\dot{\varepsilon}) = \frac{\partial \ln(m/m+1)}{\partial \ln \dot{\varepsilon}} + m > 0$$

## Results and Discussion

### Characterization of Fly-ash

The reinforcement fly ash grade P60 was procured from M/s Drik India,Nasik and used in the development of AA fly-ash composite. The 5 and 10 wt.% fly ash is added in the composite to improve its mechanical properties.

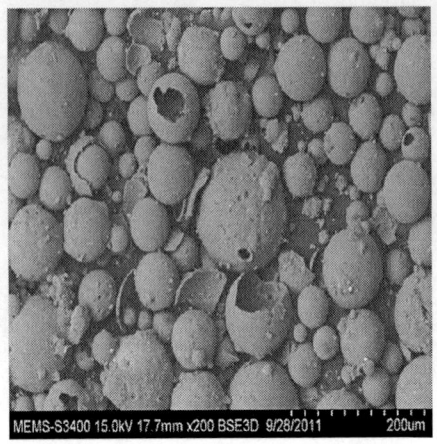

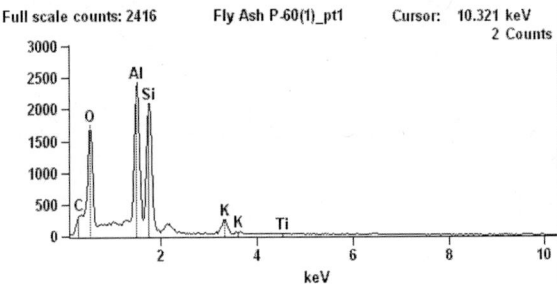

Figure 5.    SEM and EDAX analysis of Fly ash particles

## XRD Analysis of fly ash

The mineralogical structure of Fly ash was determined using X-ray diffraction. Figure . 6 shows the main mineral phases presented in the Fly ash as determined by means of XRD. The XRD patterns of the fly ash sample has peaks characteristic of quartz ($SiO_2$), lime(CaO), mullite($Al_6Si_2O_{13}$), alumina($Al_2O_3$) and iron oxides such as magnetite($Fe_3O_4$) and hematite($Fe_2O_3$) which occur in crystalline form. However, the most common phases and minerals found in these ash samples include quartz and mullite. Quartz may be considered as the primary mineral present in the Fly ash sample and indicated by sharp peaks in the diffraction patterns. In this ash sample, the most intense peak near $2\theta=26.66°$ is identified as the main peak due to quartz (101). Lime (CaO) mineral is also present in the Fly ash sample and the peak near $2\theta=26.40°$ is identified as lime. Along with the lime mineral, the concurrence of strong peaks close to $2\theta=21.0°$ indicates quartz. The presence of heavy minerals like magnetite and hematite are indicated by their respective peaks near $2\theta=35.4°$ and $2\theta=33.2°$. The peak at $37.2°$ shows the presence of $Al_2O_3$

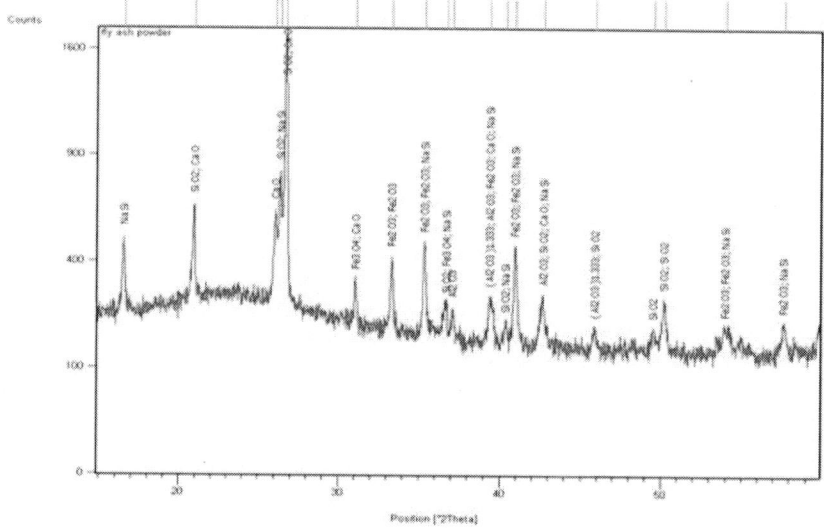

**Figure 6.** X-Ray Diffraction Pattern of Fly ash

**Microstructure of fly ash composite**

Metallographic analysis of the composites was carried out in as extruded and ECAP condition using optical microscope and in SEM. The cut samples were polished sequentially on 220, 400, 600, 800 and 1000 grit silicon carbide papers and diamond polished on satin cloth. The samples were etched using freshly prepared Keller's reagent (for 30 s) and washed in running tap water and dried using a below drier. The microstructure extruded Aa6061 base alloy and 5 wt % and 10wt% aluminum fly ash composites are shown in Fig.7 (a) (b) and (c) respectively and Figure.8

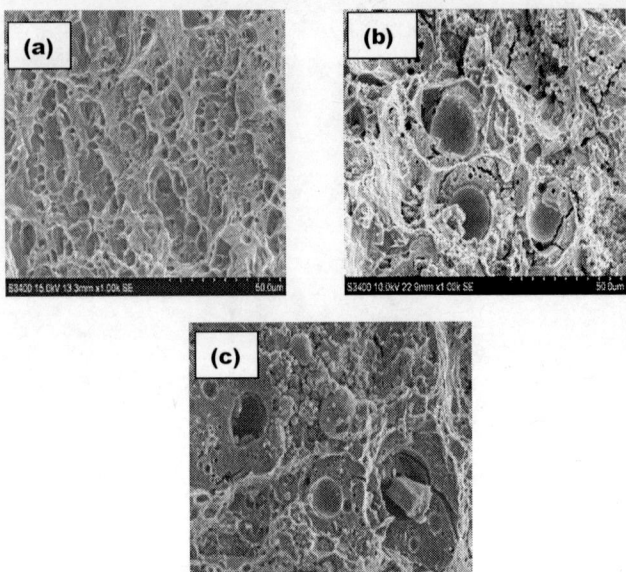

**Figure.7** Fractographs showing the
(a) AA6061 base alloy, and
(b) AA6061-5% FA composites
(c) AA6061-10% FA composites

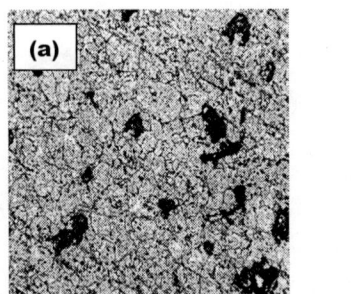 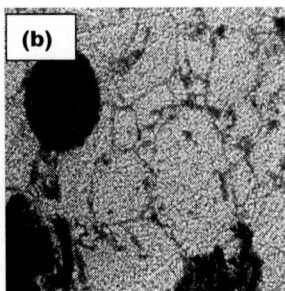

**Figure 7.** Microstructure of ECAP composite

## Mechanical properties

The mechanical properties such as ultimate tensile strength, 0.2% yield strength and percentage elongation have been evaluated for AA 6061 base alloy and fly ash composites and are listed in Table 2 and shown in Fig.7. The hardness of the base alloy and composite listed in Table 3.

Table 2.    Tensile test results of AA6061 base alloy and AA6061/fly ash composites

| Condition (Rolled) | 0.2% Y.S (MPa) | UTS (MPa) | % Elongation |
|---|---|---|---|
| AA6061 + 0% wt. fly ash | 122 | 184 | 10 |
| AA 6061 + 5% wt.fly ash | 136 | 222 | 3.45 |
| AA 6061 + 10% wt. fly ash | 184 | 249 | 3.20 |

## Hardness

The hardness have been evaluated for AA 6061 base alloy and fly ash composites with Leco Vickers Micro hardness tester and the hardness values of the base alloy and composites are listed in Table 3. The hardness of the aluminum fly ash composites increase with the fly ash reinforcement after ECAP. The SEM micrograph in Fig.7 and Fig.8 shows uniform distribution of the fly ash reinforcement in the aluminum matrix. The increase in the micro hardness is due strain fields created around fly ash particles because of the difference in the thermal expansion coefficients of aluminum base alloy and fly ash particles and after secondary processing. The strain fields piles up dislocations and the interaction between dislocations and fly ash particles offer resistance to the propagation of cracks. The grain refinement provided by the fly ash particles during solidification and after ECAP is responsible for increase in the micro hardness.

Table 3 Hardness of AA6061 base alloy and fly ash composites

| Grade | Hardness ( VPN) |
|---|---|
| AA6061   base alloy | 48 |
| AA 6061 + 5% Fly ash | 52 |
| AA 6061 + 10% Fly ash | 61 |
| AA6061  + 20% Fly ash | 70 |
| AA6061  ECAP II Pass | 75.6 |
| AA6061 -10%FA ECAP II Pass | 88.6 |

## Hot compression testing of cast 6061 -10%FA composite

Typical true stress-true plastic strain curves for the 6061 -10% Fly ash composite and the base alloy recorded at 350°C and different strain rates are shown in Fig.8 and Fig.9. The curves at temperatures lower than 200 and 500°C are similar to those in Figure 1 and exhibit continuous flow softening at all strain rates as listed in Table.4 and Table.5

**Table 4.** Value of flow stress in (MPa) of 6061 base alloy at different temperature and different strain rates for various strains.

| Temperature °C | True strain | Strain rate | | |
|---|---|---|---|---|
| | | 0.01 | 0.1 | 0.5 |
| 200 | 0.10 | 21.1234 | 24.0009 | 24.0182 |
| | 0.20 | 25.1532 | 28.8033 | 27.4627 |
| | 0.30 | 26.9968 | 30.2700 | 28.5132 |
| | 0.40 | 28.2381 | 30.8948 | 28.9993 |
| | 0.50 | 29.1055 | 31.2235 | 29.1995 |
| 350 | 0.10 | 17.0690 | 15.6920 | 9.9899 |
| | 0.20 | 17.6046 | 16.4894 | 15.8330 |
| | 0.30 | 17.5082 | 16.4438 | 15.3241 |
| | 0.40 | 17.4168 | 16.4009 | 14.8604 |
| | 0.50 | 17.2521 | 16.3066 | 14.4593 |
| 500 | 0.10 | 6.1204 | 8.0666 | 8.7596 |
| | 0.20 | 6.0420 | 8.1273 | 8.8732 |
| | 0.30 | 5.9451 | 8.1923 | 8.9669 |
| | 0.40 | 5.8573 | 8.2500 | 9.0551 |
| | 0.50 | 5.7678 | 8.2986 | 9.0797 |

Table 5    Value of flow stress in (MPa) of 6061-10%FA composite at different temperature and different strain rates for various strains

| Temperature °C | True strain | Strain rate | | |
|---|---|---|---|---|
| | | 0.01 | 0.1 | 0.5 |
| 200 | 0.10 | 15.3435 | 16.6219 | 16.9356 |
| | 0.20 | 18.0720 | 19.0775 | 19.4028 |
| | 0.30 | 19.1647 | 20.1071 | 20.5782 |
| | 0.40 | 19.7899 | 20.6956 | 21.2863 |
| | 0.50 | 20.2012 | 21.1097 | 21.7180 |
| 350 | 0.10 | 12.9011 | 11.9058 | 11.9058 |
| | 0.20 | 13.4260 | 12.7031 | 12.7031 |
| | 0.30 | 13.5944 | 13.0340 | 13.0340 |
| | 0.40 | 13.6222 | 13.2404 | 13.2404 |
| | 0.50 | 13.5955 | 13.3538 | 13.3538 |
| 500 | 0.10 | 6.2863 | 6.7162 | 6.1823 |
| | 0.20 | 6.3050 | 6.9213 | 6.5433 |
| | 0.30 | 6.1884 | 6.9326 | 6.6348 |
| | 0.40 | 6.0609 | 6.9467 | 6.6743 |
| | 0.50 | 5.9520 | 7.0043 | 6.7052 |

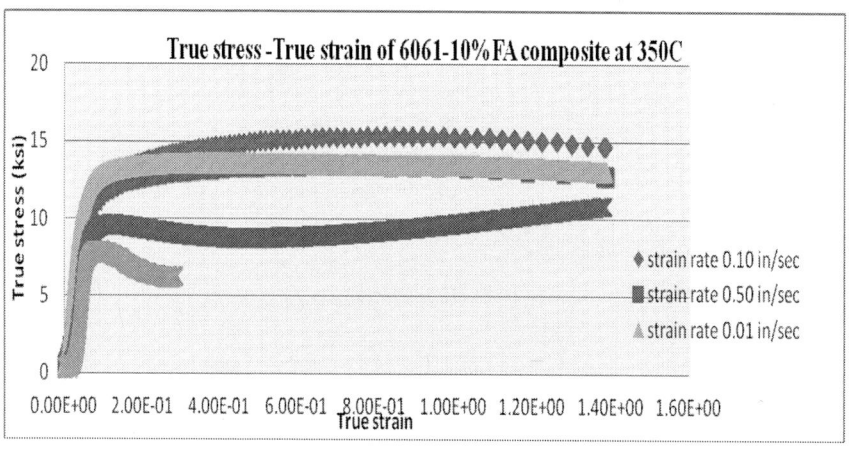

Figure 8.    True stress True strain plot of AA6061 -10 % FA composite

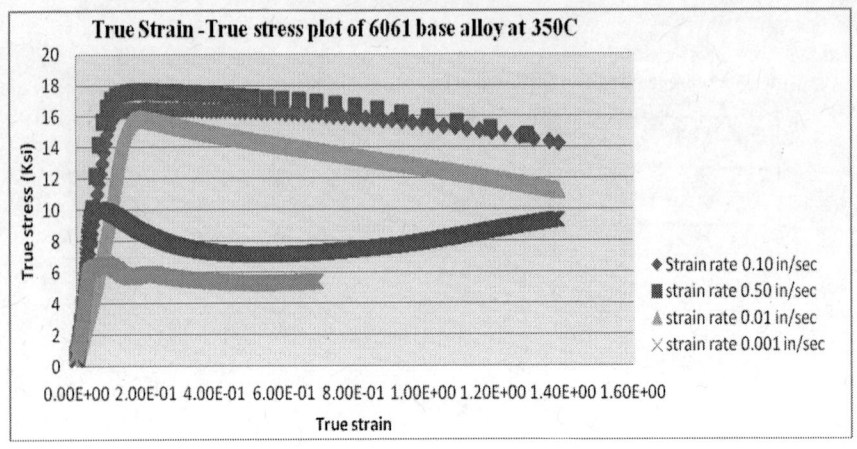

Figure 9.  True stress True strain plot of AA6061 base alloy

## Strain rate sensitivity

The strain rate sensitivity of aluminum alloy AA6061 and aluminium fly ash composite is shown in Fig.10 and Fig.11. It is found that the strain rate varies with the increase in the temperature for both AA6061 base alloy and AA6061-10% Fly ash composites

## Stain rate sensitivity of AA6061 base alloy

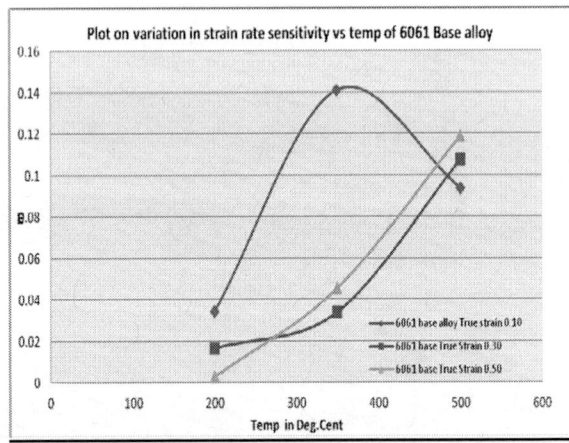

Figure 10.  Plot on variation in strain rate sensitivity of AA6061 base alloy

**Stain rate sensitivity of AA6061-10%FA**

Figure 11.  Plot on variation in strain rate sensitivity of AA6061-10%FA composite

## Processing Maps

Two compression tests for AA6061 and AA6061-10% FA composites were performed to validate results obtained from simulations performed in DEFORM 3D. Rest of the results was obtained from the simulations by varying strain rate and temperature. Data interpretation and processing was done in MATLAB.

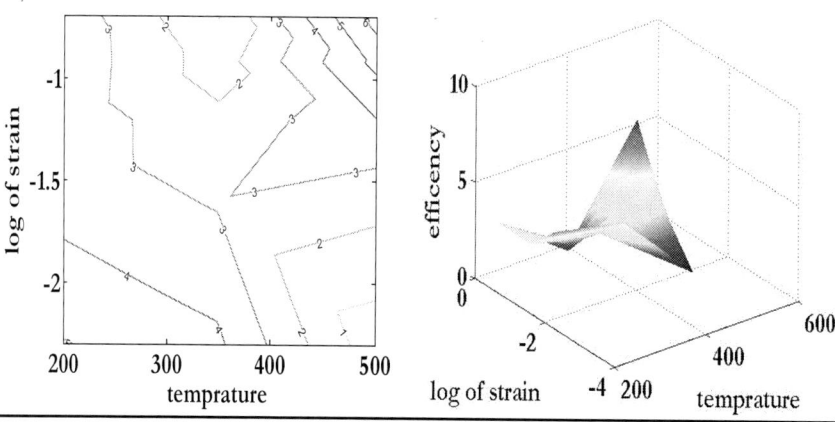

Figure 12.  Deformation maps of AA6061 stir casted base alloy

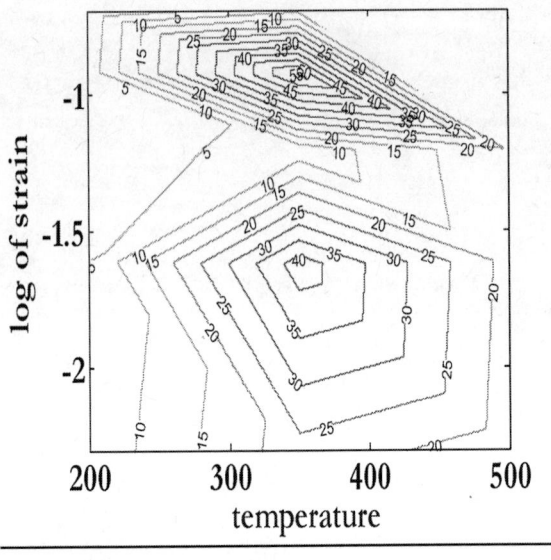

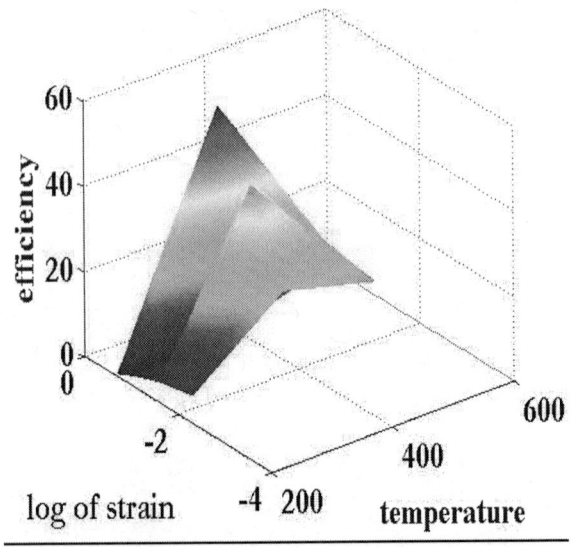

Figure 13.   Deformation maps of AA6061-10%FA composite

## Conclusions

Metal matrix composite of Al 6061 containing 10% Fly ash was synthesized successfully by using liquid metal stir casting method. Microstructure of as cast composites revealed agglomeration of fly-ash particles and interfacial bond between matrix and fly ash particles is weak. The hardness of MMC increases with increase in fly ash content and after secondary processing such as extrusion and ECAP due to grain refinement and induced strain fields near the reinforcement particles. The ultimate tensile strength, Yield strength of the composite increases with the fly ash reinforcement and ductility decreased with increase in Fly ash content. Equal channel angular pressing (ECAP) led to grain refinement and breakage of fly ash agglomerates in case of composites leading to near uniform distribution of fly ash particles. The deformation mechanism maps shows the different domains with specific micro structural mechanisms and the "safe" mechanisms were chosen for processing of aluminum fly ash metal matrix composites.

## References

1. V.K. Lindroos, M.J. Talvitie, "Recent advances in metal matrix composites", *Journal of Materials Processing Technology*, 53(1995) pp 273-284

2. E.A. Starke Jr., J.T. Staley, "Application of modern aluminium alloys to aircraft", Prog. *Aerospace Science*, Vol.32, pp 131-172, 1996

3. K.A. Lucas, H.Clarke, "Corrosion of Aluminium-based Metal Matrix Composites", Research Studies Press Ltd, 1993

4. J.P.Immarigeon, R.T. Holt, A.K. Koul, L. Zhao, W. Wallace, & J.C. Beddoes, "Lightwight materials for aircraft applications", *Materials Characterisation*, 35(1995), pp 41-67

5. I.J.Polmear, "Light alloys : From traditional alloys to nanocrystals", Elsevier, Fourth Edition, 2006, pp 75, 117-141

6. E.R. Obi, "Corrosion Behaviour of Fly Ash-Reinforced Aluminum-magnesium Alloy A535 Composites", Master of Science Thesis, University of Sasketchewan, Saskatoon, Sep 2008.

7. F.L. Matthews, R.D. Rawlings, "Composite Materials: Engineering and Science", I Edition, Chapman & Hall, 1994

8. K.K.Chawla, "Composite Materials: Science and Engineering" II Edition, Springer, 1998

9. T.P.D Rajan, R.M. Pillai, B.C. Pai, K.G. Satyanarayana & P.K. Rohatgi, "Fabrication and characterization of Al-7Si-0.35Mg/Fly ash metal matrix composites processed by different stir casting routes", *Composites Science and Technology* 67 (2007) pp 3369-3377

10. S. Sarkar, S. Sen, S.C. Mishra, "Studies on Aluminium-Fly ash Composite produced by Impeller Mixing", *Journal of Reinforced Plastics and composites*, Vol 00, 2008

11. Emmanuel Gikunoo, "Effect of fly ash particles on mechanical properties and microstructure of aluminium casting alloy A535", MS Thesis, University of Sasketchewan, Ottawa, Canada, Nov 2004

12. M Ramachandra, K. Radhakrishna, "Study of abrasive Wear Behavior Of Al-Si (12%)-SiC Metal Matrix Composite Synthesized Using Vortex Method", International Symposium of Research Students on Materials Science and Engineering, IIT Madras, Dec 2004

13. J. Bienia, M. Walczak, B. Surowska, J. Sobczak, "Microstructure And Corrosion Behaviour Of Aluminum Fly Ash Composites", *Journal of Optoelectronics and Advanced Materials* Vol. 5, No. 2, June 2003, pp. 493-502

14. S. Naher, D. Brabazon, L. Looney, "Simulation of the stir casting process", *Journal of Materials Processing Technology* Volumes 143-144, 20 December 2003, Pages 567-571

15. P.K. Rohatgi, R.Q. Guo, H. Iksan, E.J. Borchelt, R. Asthana, "Pressure infiltration technique for synthesis of aluminium-fly ash particulate composite" *Materials Science and Engineering*, V 244(1998), pp 22-30

16. Boq-Kong Hwu, Su-Jien Lin, Min-Ten Jahn, "Effects of Process parameters on the properties of Squeeze-case SiCp-6061 Al Metal Matrix composites", *Materials Science and Engineering* A, 207(1996), pp 135-141

17. Jerzy Sobczak, P. Darlak, R.M. Purgert, N. Sobczak, A. Wojciechowski, "The technological aspects of ALFA Composites Synthesis", Solidification Processing of Metal Matrix Composites – Rohatgi Honorary Symposium, Edited by N. Gupta, and W.H. Hunt TMS (The Minerals, Metals & Materials Society), 2006

18. V Amigo, J.L. Ortiz, M.D. Salvador, " Microstructure and mechanical behaviour of 6061 Al reinforced with silicon nitride particles, processed by powder metallurgy", *Scripta Materialia*, 42(2000), pp 383-388

19. H.S. Jailani, A. Rajadurai, B. Mohan, A.S. Kumar, T. Sornakumar, " Multi-response optimization of sintering parameters of Al-Si alloy/fly ash composite using Taguchi method and grey relational analysis", *International Journal of advanced Manufacturing Technology*, 2009

20. R.Q. Guo, P.K. Rohatgi, "Chemical reaction between aluminium and fly ash during synthesis and reheating of Al-fly ash composite", *Metallurgical and Materials Transactions* B, Vol. 29B, June 1998, pp 519-525

21. R.Z. Valiev, T.G. Langdon, "Principles of equal-channel angular pressing as a processing tool for grain refinement", *Progress in Materials Science* 51 (2006) 881–981

22. M. Furukawa, Z. Horita, T.G. Langdon, "Processing by Equal-Channel Angular Pressing: Applications to Grain Boundary Engineering", *Journal of Material Science*, Vol. 40, No. 4, pp. 909-917.

23. K. Nakashima, Z. Horita, M. Nemoto, T. G. Langdon, "Influence of channel angle on the development of ultrafine grains in equal-channel angular pressing", *Acta Materialia*, Vol. 46 (1998), pp 1589-1599.

Advanced Composites for Aerospace, Marine, and Land Applications
Edited by: Tomoko Sano, T.S. Srivatsan, and Michael W. Peretti
TMS (The Minerals, Metals & Materials Society), 2014

# EFFECT OF COMPOSITION OF B₄C-ALUMINUM COMPOSITES ON MECHANICAL PROPERTIES AND RESISTANCE CORROSION

Lucio Vázquez[1], Edgardo Hernández[1], Alejandro Altamirano[1], Víctor Cortés[1], Elizabeth Garfias[1], Elizabeth Refugio[1] and Manuel Vite[2]

[1] Universidad Autónoma Metropolitana-Azcapotzalco,
Av. San Pablo 180, Col. Reynosa Tamaulipas México D.F., C.P. 02200, **México**

[2] **Instituto Politécnico Nacional**, Unidad Profesional Adolfo López Mateos,
Building 5, Third Floor, Col. Lindavista, México D.F., C.P. 07738, **México**

## Abstract

Purposes of this work were to define the compositions which allowed removal of specimens from the die after pressing without crumbling, the only compositions which made this possible were those containing 3%, 5% and 7% in weight of B₄C (the rest was aluminum), and to determine the effect that B4C content has on the mechanical properties of the composites, resistance to corrosion, and examination by optical and scanning microscopies. Samples were prepared by a P/M technique. Compression properties were better in the composites than in aluminum and improved with carbide content. Samples of all the experimental compositions subjected to salty steam within a salty steam chamber did not show any corrosion. Optical microscopy revealed that Al and the composites have a like cell structure. Scanning microscopy revealed that morphology of the B₄C particles is polyhedral. There were identified also particles of SiC and Al₂O₃, coming probably from the mill.

Keywords:    Composites, B₄C-Al, Mechanical, Properties, Corrosion

## Introduction

In the last years the metal composites (MMCs) have attracted growing interest of the automotive aeronautical, electronic, and space industries [1, 2]. These materials are alluring because combine high values of ratio strength/density, wearing resistance, stiffness, good thermal conductivity and stability, as well as an excellent formability which enables the production of complicated forms. Their applications to make high technology devices justify that these materials be considered as advanced materials [3].

The most studied MMCs are matrix aluminum with ceramics as reinforcement and among the latter the most spread used are Al2O3 and SiC because their densities are similar to that of aluminum. Use of B4C is promising because the high resistance to wearing and density slightly lower that of aluminum. The composites fabrication method simpler is by powder metallurgy: it consists of milling the mixture of aluminum and ceramic, pouring the mixture into a die, pressing it, drawing the composite piece from the die and sintering it in a furnace at an appropriate temperature with controlled atmosphere.

## Experimental Procedure

### Manufacture of samples

Experimental Materials. Aluminum used consists of granules, 100% passes through a 50 meshes sieve and are completely retained by a 100 meshes sieve. The chemical composition of the powder aluminum is shown on Table 1.

Table 1.  Chemical Composition of Aluminum

| Element | Sb | Ba | Cd | Pb | Cr |
|---|---|---|---|---|---|
| Concentration (ppm) | 18 | 40 | 8 | 9 | 5 |

Rest to 100%: aluminum B4C was as a powder, size 1200 meshes sieve. Purity 99.95 %..
Weighting of Materials. Components of the composite were weighted in an analytical balance.

### Grinding.

The ratio of load of balls to B4C-Al mixture was 10:1. The load composition consisted of 900 g of zirconia and alumina one cm diameter spheres (50% of each material). The grinding parameters used were mill speed 340 rpm, grinding time 12 h.

### Pressing.

The mixture B4C-Al was poured into D2 steel dies to make specimens of rounded or rectangular transverse section, It was applied a 300MPa pressure at environmental temperature, sustaining the load for ten minutes. In this way there were made circulat specimens of 2 cm diameter and 2, 4 and 6 cm heights; rectangular specimens 0.8 cm x 2.54 cm transverse section and 5.08 cm height. A thin film of SAE 20W-50 Quaker State oil was applied to the wall dies to reduce friction, to avoid sticking of the material to the walls as well as to circumvent crumbling of the specimen during the extraction operation.

### Sintering.

The specimens were sintered in a furnace using a nitrogen atmosphere with a flow pressure of 1MPa. The thermal cycle consisted of 3 h at 550°C. After this time furnace was off, nitrogen flow was interrupted, samples were left within the furnace for cooling until environmental temperature.

### Sampling Testing and Results

Definition of the composite composition. One of the purposes of this work was to define the maximum amount of B4C which could be incorporated to the host metal, Al, to still to get a good cohesion. The aim was reached by preparing round specimens 6 cm height with 10, 20, 30, 40, and 50% in weight of B4C. After compression following the procedure described above, the specimens crumbled while they were released from the die, Figure 1. Samples with 10% B4C displayed less crumbling. Therefore, it was decided to prepare specimens containing 3, 5 and 7% B4C.

### Density.

Density of the specimens was determined by the Archimedes method using a hydrostatic balance. Five rounded specimens 2 cm height for each composition were weighed out in a balance, then samples were weighed submerged in water and the displaced volume of water was measured. Figure 2 shows densities as a function of composition.

### Microhardness.

Five round samples of 2.0 cm height for each composition were grounded and polished to a mirror finishing. Vickers hardness was used with a 50 g load. Figure 3 shows graphs of average hardness vs. composition.

### Compression.

All the specimens tested in microhardness were used to carry out the compression tests. Figure 4 presents average values of the compression property vs. composition

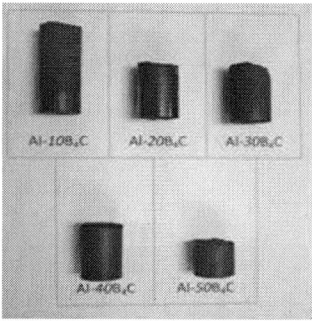

**Figure 1.** Crumbling of specimens

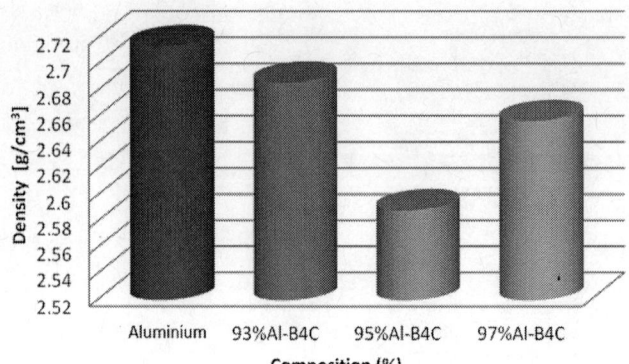
figure 2. Density vs Composition

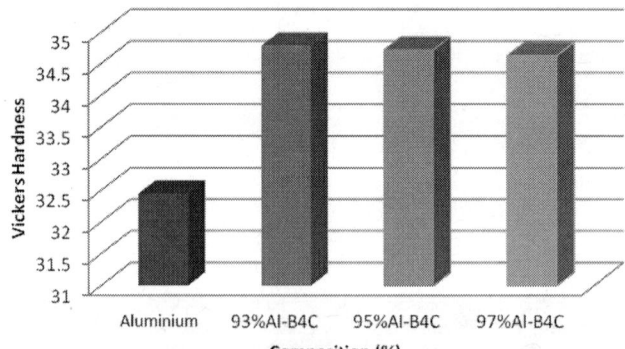
Figure 3. Microhardness vs Composition

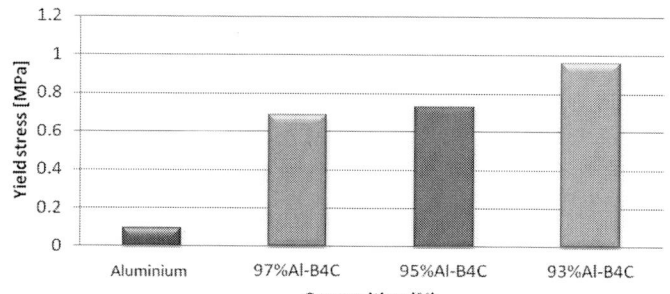

Figure 4(a). Average yield stress vs. Composition

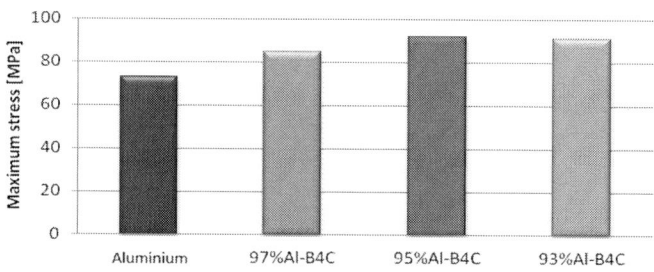

Figure 4(b). Maximum stress vs. Composition

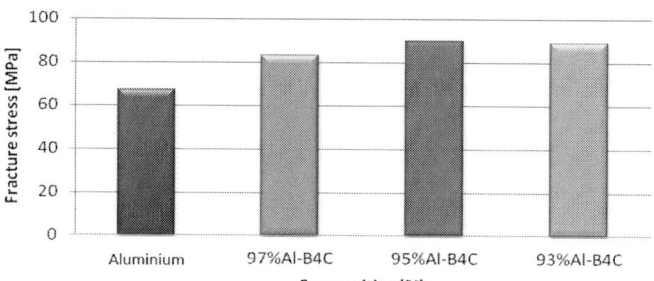

Figure 4(c) Fracture stress vs. Composition

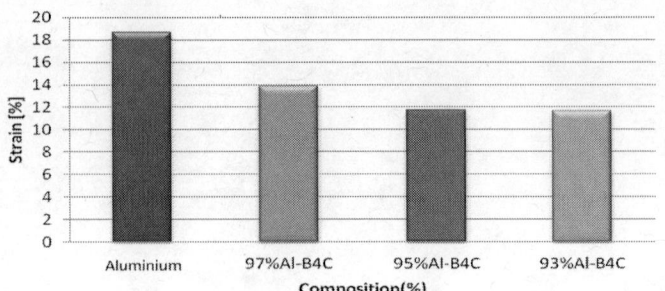

Figure 4(d), Average strain vs. Composition

## Impact Test.

There were used five 1cmx1cm square transverse section 5 cm long specimens without groove for each composition. The material was fragile therefore it was not possible to machine a groove according to ASTM standard However, the results allowed to estimate and to compare the absorbed energy by aluminum and the composites for the different experimental compositions. Figure 5 displays the compact behavior.

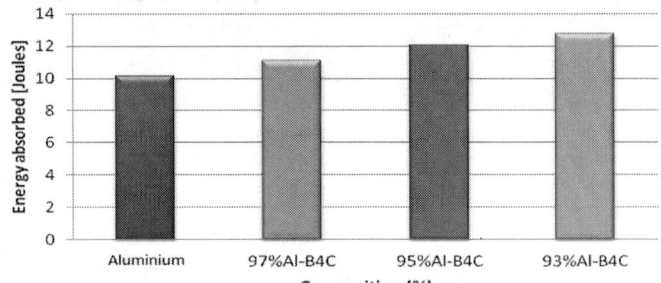

Figure 5. Energy absorbed vs Composition

## Wearing Test.

Five samples 0.8 cmx2.5 cm transverse section and 5.0 cm height for each composition were used for this test. The tests were carried following ASTM standard G-65. The wearing machine looks like a balance devise. For the tests was used 1 kg load. The specimen is fixed on the surface of the bar which hangs up on one end of the balance arm. A wheel covered with rubber rotating at 200 rpm touches the specimen as the load is applied and the abrasive powder, silica sand 0f 50 -70 meshes size, drops from a hopper, Figure 6. Figure 7 shows the results obtained for the experimental compositions.

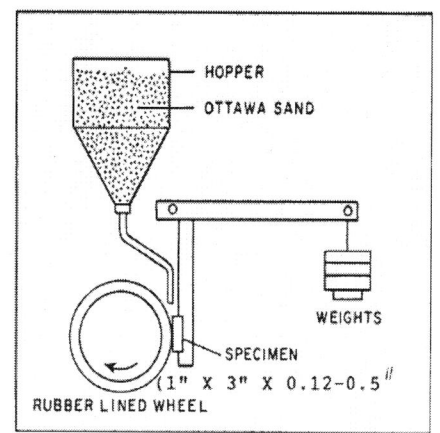

Figure 6.  Wearing machine

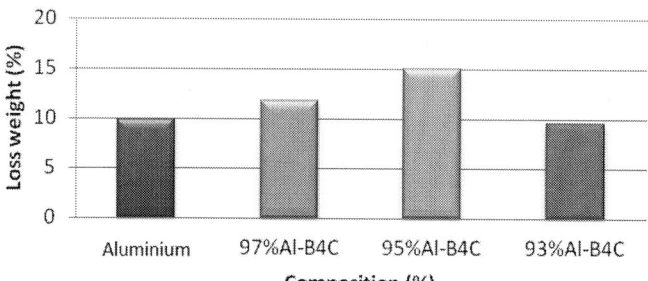

Figure 7. Loss weight (%) vs. Composition

**Corrosion Tests.**

The corrosion behavior of the specimens for all the compositions was evaluated by a salt spray test following the procedure used for paints. Samples were placed in a salt spray chamber. The salty solution had a concentration 5% in weight of NaCl and was used a water quality type IV according to ASTM D1193-08 Standard. It was intended to evaluate the degree and distribution of rust by ASTM D 602-08 Standard, but after 336 h exposition no rust was observed. Figure 8 shows the specimens after the test.

**Figure 8.** Specimens of composite 97%Al-B$_4$C, (a) before testing, (b) after test.

**Optical Microscopy.**

The material of all the specimens forms cells, characteristic which is more evident at 100X, it is attributed to the procedure used to make the specimens: by compression, Figure 9. The distribution of B4C is homogeneous, although it seems that the reinforcement concentrates on the boundaries of the cells.

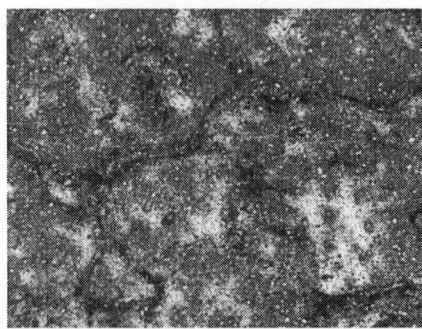

**Figure 9.** Micrograph at 100X of composite 97% Al-B$_4$C.

Scanning Electronic Microscopy (SEM). SEM renders micrographs of better quality. Figure 9 corresponds to the same sample shown in Figure 10. Figure 11 shows the distribution of B4C particles.

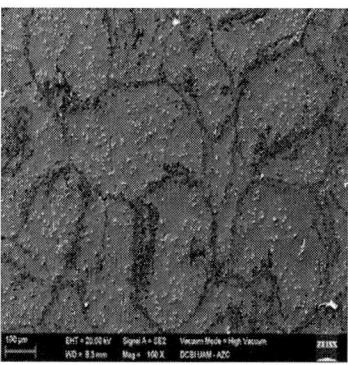

**Figure 10.**  Electron micrograph at 100X of specimen with 97% Al.

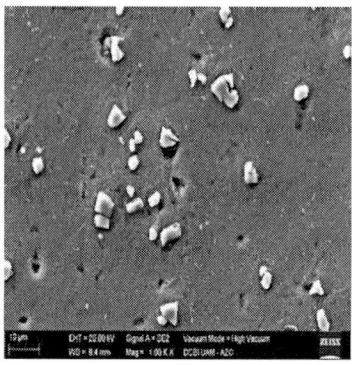

**Figure 11.**  Electron micrograph at 1000X of specimen with 97% Al-B$_4$C

## Analysis of Results

<u>Composition.</u>

Composites containing more than 10% in weight of B4C exhibit a poor cohesion between the reinforcement (B4C) and the matrix (Al) which is the cause of crumbling. Figure 1.

**Density.**

Aluminum presents a higher density than the composites. There is not observed a steady decrease tendency of density with increasing contens of B4C, Figure 2. This behavior is attributed to heterogeneity in the structure of the composites. Figure 2.

**Microhardness.**

Aluminum displays a lower hardness than the composites. Hardness increases with content of B4C. Figure 3.

**Compression Properties.**

**Yield Stress.**

This property is higher for the composites than for aluminum. It increases with B4C content, Figure 4(a).

**Maximum Stress.**

Aluminum exhibits a lesser value than the composites, and in these materials maximum stress increases with B4C content, Figure 4(b).

**Fracture Stress.**

This property follows almost the same trend than the maximum stress: lower for aluminum than for the composites and the value increases as B4C content increases, Figure 4(c).

Strain until Fracture. Strain is higher for aluminum than for the composites and in the latter materials decrease as B4C grows, Figure 4(d).

**Impact Test.**

Energy absorbed shows a trend of growing creciente from aluminum to composites with higher B4C content. Toughness is higher for the composites and increases with B4C, Figure 5.

**Wearing Test.**

Although wearing resistance is higher for the composites than for aluminum, values of loss material inside the group of composites do not follow a trend which allows to define the roll of B4C. There was observed that during the test sometimes the silica sand flow interruptes for short times, this could be reason of the variability of results.

**Corrosion Tests.**

After exposition of the samples for 14 days to the salty fog, no corrosion was observed.

**Microscopy.**

The specimens exhibit a cell structure, probably a consequence of the way of preparation by compression which is similar to extrusion, like milled meat coming from the mill. Particles of

B4C deposit on the cell boundaries, but also on the interior of the cells. Electron microscopy reveals that the B4C particles are angular and rounded and are distributed over the whole mass of aluminum.

## Conclusions

| | |
|---|---|
| Composition. | P/M technique allows preparing composites with 10% or less of B4C. Composites with higher content crumble during the release from the die. |
| Microhardness. | Hardness increases with B4C content. |
| Compression tests. | Yield, maximum and fracture stress increase with B4C content. Strain decreases as B4C content increases. |
| Impact test. | Toughness increases as B4C increases. |
| Wearing test. | Resistance to wear improves with B4C content. |
| Corrosion test. | Composite and aluminum specimens do not undergo any corrosion when exposed to salty fog. |
| Microscopy. | Boron carbide particles deposit on cell boundaries, but distribution within the cells is homogeneous. |

## References

1. Dowson G. *Introduction to powder metallurgy, the process and its products*, European Powder Metallurgy Association – Education and Training, 2000.

2. Baetz, J. G., *Metal Matrix Composites: Their Time has Come"*, November 1998, pp. 14-16.

3. Goni, L., Mitxelena, I., and Coleto, I., *Development of low cost metal matrix composites for commercial applications*, Material Science and Technology, v. 16, No. 7 – 8. July – August 2000, pp. 743 – 746.

4. Bhagat, R. B., *Advanced aluminum powder metallurgy, alloys and composites*, ASM Handbook, v.7, 1998, pp. 840.

5. Ruiz-Navas E. M., Da Costa, C. E., *Lopez* F. V., and Castello, J. M. T., *Mechanical alloying: a method to metallic powder composite materials*, Revista Metalúrgica, Madrid, v. 13, No. 4, 2000, pp. 279-286.

6. Torralba J. M., Da Costa, C. E. and Velasco F., *P/M aluminum matrix composites: an overview*, J. of materials Processing Technology, v. 133, Issues 1-2, 2003, pp. 203-206.

7. Kaczmar, I., Pietrzak, K., and Wlosinski, W., *The production and application of metal matrix composite materials*, J. of Mmaterials processing technology, v. 106, 2000, pp. 58-67.

8. Annual Book of Standards, v. 6-2, 2010.

# BACTERIAL CELLULOSE ENHANCES BETA PHASE IN PVDF

Vivek Verma[1*], P. S. M. Rajesh[2#], Sampada Bodkhe[2#], Sudhir Kamle[2]

[1]Department of Materials Science and Engineering,
Centre for Environmental Science & Engineering,
Thematic Unit of Excellence on Soft Nanofabrication
with
Application in Energy, Environment & Bio platform
**Indian Institute of Technology (IIT-Kanpur)**
Kanpur, India 208016

[2]Department of Aerospace Engineering
**Indian Institute of Technology** (IIT – Kanpur)
Kanpur, India 208016
*Corresponding author: Email: vverma@iitk.ac.in,

[#]Present address: Department of Mechanical Engineering,
**Ecole Polytechnique de Montreal**, Canada H3T 1J4

## Abstract

Polyvinylidene fluoride (PVDF) is a smart material owing to its pyro-, piezo- and ferroelectric properties. For practical application, e.g., as a sensor material however; the non-polar, low-energy, α-phase of PVDF must be transformed into polar β-phase. Bacterial cellulose was used to study its effect on α to β-phase transformation in PVDF. Bacterial cellulose produced from *Acetobacter xylinum* bacteria was incorporated (0.5, 1 and 2% by weight) into PVDF films using solution casting technique. While ultrasonication provided energy for phase transformation, the filler helped retain the β-phase. The evolution of this phase was confirmed and estimated using FTIR and XRD studies.

## Introduction

The ferroelectric, dielectric and piezoelectric properties of poly (vinylidene fluoride) (PVDF) are highly sought after for sensing[1, 2] and actuation[3, 4]. Being semicrystalline, PVDF has both amorphous and crystalline regions. In crystalline regions, there are five phases of PVDF reported in literature, namely α, β, γ, δ and ε. PVDF finds its application in electronic materials due to its polar β-phase. The low energy α-phase however, is energetically more stable than polar β-phase at room temperature, hence, is the major phase in PVDF. Thus, for employment of PVDF as an electronic material, it is imperative to transform α-phase to β-phase. Several techniques including ultrasonication[5], annealing[6], poling[7], vapor deposition[8], stretching[9] etcetera have been used to achieve this transformation. Addition of small quantity of fillers into PVDF has also been effective to enhance β-phase. To this end, addition of carbon nanotubes[10], intercalated clay[11], barium titanate[12], etcetera have been reported in literature. Their addition not only enhances the requisite β-phase for smart application of PVDF but also can improve mechanical and thermal properties of PVDF.

We report bacterial cellulose as filler in PVDF for enhancing β-phase. Bacterial cellulose is produced by *Acetobacter xylinum*[13] bacteria. The chemical structure of bacterial cellulose is the same as that of plant cellulose but has higher crystallinity. In our previous work[14], we showed effect of various physicochemical modifications of plant cellulose on β-phase development of PVDF. It was proposed that surface charges and crystallinity of cellulose play an important role in enhancing β-phase. Continuing on this line, bacterial cellulose was chosen since it has higher crystallinity and is purer than plant cellulose. Composite blends of PVDF with 0.5, 1 and 2% (w/w) bacterial cellulose were prepared, ultrasonicated and cast into thin standing films. While ultrasonication provided energy for α to β transformation, the filler (bacterial cellulose in this case) deterred transformed β-phase from going back to α-phase. The films were characterized using X-ray diffraction (XRD) and Fourier transform infrared (FTIR) spectroscopy for phase evolution.

## Materials and Methods

### Materials

Poly (vinylidene fluoride) ($M_W$ ~ 534000), N-dimethyl formamide (DMF), dimethyl sulfoxide (DMSO) and corn steep liquor were purchased from Sigma-Aldrich; Fructose from SD fine chemical ltd.; sodium alginate and magnesium sulfate from Loba Chemie; and mono potassium phosphate, calcium chloride and sodium hydroxide from Fisher Scientific. Lyophilized *Acetobacter xylinum* were received as a generous gift from Agrilife (India).

### Production of Bacterial Cellulose

Cellulose was prepared from *Acetobacter xylinum* in 500 ml corn steep liquor (CSL)-fructose media with the following composition - 5% w/v fructose, 2% v/v corn steep liquor, 0.04% w/v sodium alginate, 0.5g $KH_2PO_4$, 0.75 mg $CaCl_2$, 0.2 mg $MgSO_4$ at 30 °C while shaking at 75 rpm for 5 days. After 5 days cellulose was removed from the media and heated at 70°C in 4% w/v NaOH for an hour. It was followed by washing the cellulose with DI water until the pH turned neutral. Bacterial cellulose was lyophilized and stored at 4 °C in a sealed petri dish.

*Film Preparation*

DMF, as solvent and acetone as antisolvent were used to solution cast PVDF films[15]. To allow β-phase formation at lower temperatures (~30 °C), DMSO (45 g/L) was used[9]. The method to prepare PVDF and PVDF composite films was followed as per the procedure employed by Benz et al[9]. To prepare PVDF-bacterial cellulose composite films, cellulose was ultrasonicated in acetone for 3 hours at a frequency/power of 33 kHz/70 W (max). DMF (10% v/v) was added to this suspension followed by 100 g/L PVDF and 45 g/L DMSO. The blend was ultrasonicated for 20 minutes while maintaining the temperature below 30 °C. After leaving the PVDF-bacetrial cellulose blend at 40 °C for 20 minutes, the films were cast on flat polyester sheet. The concentration of bacterial cellulose in PVDF composite films was 0.5, 1 and 2% (w/w). The thickness of the films lied between 300-500 μm. For Fourier transform infrared (FTIR) spectroscopy, thinner films of ~30-60 μm were cast.

*Characterization*

X-Ray diffraction was used to scan the samples from 16° to 28° with a BRUKER D8 Focus instrument at room temperature with Cu target, $K_\alpha$ radiation at 40 kV and 40 mA. A slow scan rate of 0.2 °/min was used to minimize the noise in the polymer films. The FTIR spectra of films were obtained from 400-4000 cm$^{-1}$ using Bruker Vertex 70 instrument. However, spectra between 400-1000 cm$^{-1}$ are depicted here for clarity. Ziess EVO-50 scanning electron microscope (SEM) was used to characterize structure of bacterial cellulose. Accelerating voltage of 15 kV was applied.

## Results and Discussion

The SEM image of bacterial cellulose shows high aspect ratio fibers (figure 1). The cellulose fibers had similar structure as reported in literature[16].

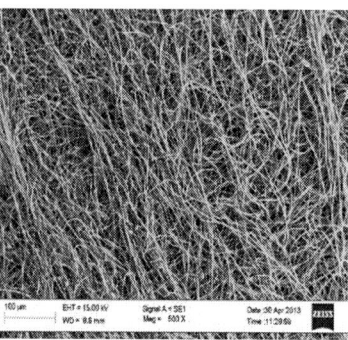

**Figure 1.** Scanning electron micrograph of freeze dried cellulose film produced by Acetobacter xylinum

In X-ray diffractograms (figure 2), the peaks at 2θ = 18.42°, 19.9° and 26.5°, corresponding to α-phase, are prominent in the neat PVDF sample, which agrees well with those reported in references[5, 17, 18]. However, the peaks in PVDF-filler composite films present a minor shift from those in blank PVDF. The presence of β-phase is characterized by (i) diminishing α reflections [10, 19] and (ii) peak broadening especially of the α-peak at 19.9°[20]. The peak intensities at 18.42° and

26.5° have been reduced. In addition to that, the peak at 19.9° has widened and shoulders have appeared towards the higher angle side. All these observations serve as an evidence that β-phase exists in the samples containing fillers.

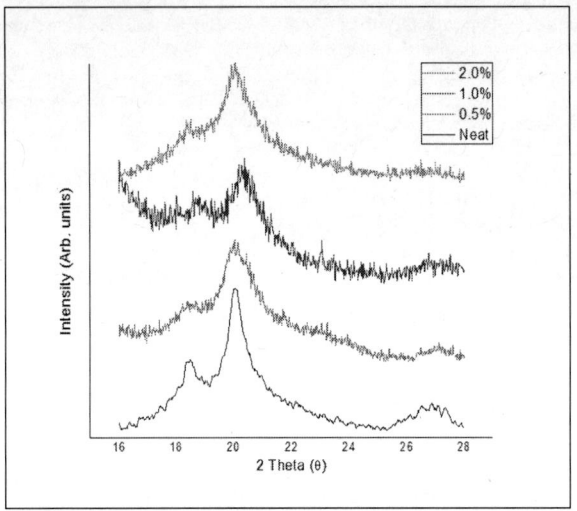

**Figure 2**: X-ray diffractograms of bacterial cellulose PVDF composites

For FTIR spectroscopy, literature indicates that the spectral bands around 490, 530, 615, 766, 795 and 974 $cm^{-1}$ correspond to the α-phase of PVDF while the ones around 512 and 840 $cm^{-1}$ correspond to the presence of β-phase[5, 20, 21]. The IR spectra of the various PVDF films with α and β-phases are given in Figure 3.

In our study, we applied Beer-Lambert law to FTIR data to perform a quantitative analysis of the phases present in the PVDF composites. This law relates the absorbance to the concentration of the absorbing species. The fraction of β-phase present in the films was calculated using the formula:

$$\text{Fraction of β-phase, } F_\beta = \frac{A_\beta}{1.26(A_\alpha) + A_\beta} \quad (22)$$

where, $A_\alpha$ and $A_\beta$ are the absorption fractions of α and β-phase respectively and the value 1.26 is the sum of the absorption coefficients of the two phases. The absorbance at 763 $cm^{-1}$ ($CF_2$ bending and skeletal bending) was chosen as $A_\alpha$ and the maximum absorbance around 838 $cm^{-1}$ ($CH_2$ rocking) as $A_\beta$. The calculations based on the above formula led to results which are presented graphically in figure 4 and are clearly observed to be in accordance with the spectra.

The neat PVDF, as per our calculations, has 33% β-phase present in it (as indicated using a hashed line) that increases with the addition of the fillers. The enhancement in β-phase was maximum (64%) for 0.5% cellulose addition. PVDF/cellulose composites showed a constant decreasing trend: 0.5% addition showing the highest β-phase to while 2% had 57.7% of β-phase, which is still higher than the β-phase content in neat PVDF. We speculate that poor dispersion of fillers at higher weight fractions could be the reason for decline in β-phase, which decreases the number of available surface groups for interaction.

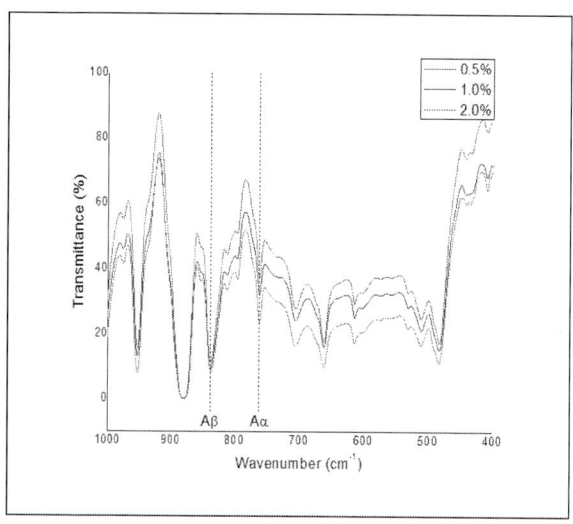

**Figure 3:** FTIR spectra of bacterial cellulose PVDF composites

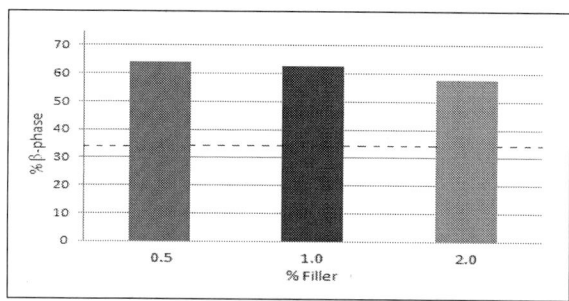

**Figure 4:** Percentage β-phase obtained at various filler ratios. The percentage of β-phase in neat PVDF is indicated via a hashed line at 33 percent.

## Conclusions

Bacterial cellulose produced from *Acetobacter xylinum* was used as a filler to improve the β-phase content in PVDF. The solution cast PVDF films contained 0.5, 1 and 2 % (w/w) bacterial cellulose. Enhancement in β-phase was confirmed using XRD and FTIR studies. The improvement in β-phase decreased with increasing cellulose content, which can be attributedto poor dispersion of cellulose at higher loading ratios. Nonetheless, with close to 100% increase in β-phase over the neat variant, bacterial cellulose PVDF composites provide an ecologically viable route to smart material synthesis.

## Acknowledgements

The authors thank Sankalp Verma, department of material science and engineering, IIT Kanpur for providing the bacterial cellulose used in this study. XRD and FTIR studies were respectively carried out at the department of materials science and engineering and the department of chemistry, IIT Kanpur. The authors thank Agrilife (India) for gifting lyophilized *Acetobacter xylinum* cells. This work was supported by the national programme on micro-air vehicles, aeronautics R&D board, DRDO, India.

# References

1. Jingang, Y.; Hong, L. *Sensors Journal, IEEE* **2008**, 8, (4), 384-391.

2. Rathod, V. T.; Mahapatra, D. R.; Jain, A.; Gayathri, A. *Sensors and Actuators A: Physical* **2010**, 163, (1), 164-171.

3. Ya-hong, Z.; Shi-lin, X.; Xi-nong, Z. In *Actuating characteristic of laminated PVDF actuators used on a beam with large deformation*, Piezoelectricity, Acoustic Waves, and Device Applications, 2008. SPAWDA 2008. Symposium on, 5-8 Dec. 2008, 2008; 2008; pp 284-288.

4. Ya-hong, Z.; Xi-nong, Z. In *Investigation of actuating characteristics of laminated PVDF actuator used on paraboloidal shells*, Piezoelectricity, Acoustic Waves and Device Applications (SPAWDA), 2010 Symposium on, 10-13 Dec. 2010, 2010; 2010; pp 128-133.

5. Yu, S.; Zheng, W.; Yu, W.; Zhang, Y.; Jiang, Q.; Zhao, Z. *Macromolecules* **2009**, 42, (22), 8870-8874.

6. Kang, S. J.; Park, Y. J.; Sung, J.; Jo, P. S.; Park, C.; Kim, K. J.; Cho, B. O. *Applied Physics Letters* **2008**, 92, (1), 012921-3.

7. Jiang, Y.; Ye, Y.; Yu, J.; Wu, Z.; Li, W.; Xu, J.; Xie, G. *Polymer Engineering & Science* **2007**, 47, (9), 1344-1350.

8. Takeno, A.; Okui, N.; Kitoh, T.; Muraoka, M.; Umemoto, S.; Sakai, T. *Thin Solid Films* **1991**, 202, (2), 205-211.

9. Benz, M.; Euler, W. B.; Gregory, O. J. *Macromolecules* **2002**, 35, (7), 2682-2688.

10. Huang, X.; Jiang, P.; Kim, C.; Liu, F.; Yin, Y. *European Polymer Journal* **2009**, 45, (2), 377-386.

11. Peng, Q.-Y.; Cong, P.-H.; Liu, X.-J.; Liu, T.-X.; Huang, S.; Li, T.-S. *Wear* **2009**, 266, (7–8), 713-720.

12. Verónica Corral-Flores, J. J. P.-H., Enrique Torres-Moye, Jorge Romero-García, Darío Bueno-Baqués, Ronald F. Ziolo. *Materials Science Forum* **2010**, 644, 33-37.

13. Hestrin, S.; Schramm, M. *Biochem J* **1954**, 58, (2), 345-52.

14. Rajesh, P. S. M.; Bodkhe, S.; Kamle, S.; Verma, V. *Electronic Materials Letters* **2013**, (Accepted).

15. Williams III, R. O.; Watts, A. B.; Miller, D. A., *Formulating Poorly Water Soluble Drugs*. Springer: London, 2012; Vol. 3.

16. Svensson, A.; Nicklasson, E.; Harrah, T.; Panilaitis, B.; Kaplan, D. L.; Brittberg, M.; Gatenholm, P. *Biomaterials* **2005,** 26, (4), 419-431.

17. Kim, G. H.; Hong, S. M.; Seo, Y. *Physical Chemistry Chemical Physics* **2009,** 11, (44).

18. Mohammadi, B.; Yousefi, A. A.; Bellah, S. M. *Polymer Testing* **2007,** 26, (1), 42-50.

19. Mago, G.; Kalyon, D. M.; Fisher, F. T. *Journal of Nanomaterials* **2008,** 2008.

20. Huang, S.; Yee, W. A.; Tjiu, W. C.; Liu, Y.; Kotaki, M.; Boey, Y. C. F.; Ma, J.; Liu, T.; Lu, X. *Langmuir* **2008,** 24, (23), 13621-13626.

21. Shah, D.; Maiti, P.; Gunn, E.; Schmidt, D. F.; Jiang, D. D.; Batt, C. A.; Giannelis, E. P. *Advanced Materials* **2004,** 16, (14), 1173-1177.

22. R. Gregorio, J.; Nociti, N. C. P. d. S. *Journal of Physics D: Applied Physics* **1995,** 28, (2), 432.

Advanced Composites for Aerospace, Marine, and Land Applications
Edited by: Tomoko Sano, T.S. Srivatsan, and Michael W. Peretti
TMS (The Minerals, Metals & Materials Society), 2014

# GEOPOLYMER FROM INDUSTRIAL WASTES: A CONSTRUCTION MATERIAL FOR 22nd CENTURY

Pradeep Kumar Rana[1*], Radha Raman Dash[2] and Ratan Indu Ganguly[3]

[1] Assistant Professor
[2] Dean (R & D)
[3] Professor & Head
Department of Metallurgical and Materials Engineering
GIET, Gunupur

## Abstract

Portland cement (PC) is used as the binder to produce concrete. The environmental issues associated with the production of PC is the release of $CO_2$ in the order of one ton for every ton of PC produced. In this work, source materials like industrial wastes such as low carbon and low calcium flyash, Redmud from alumina refinery and slag from AOD furnace of stainless steel making are used to bind the loose, coarse, fine aggregates and other un-reacted materials together to form the geopolymer concrete. The silicon and the aluminium present in the source materials reacts with an alkaline liquid, namely, $Na_2SiO_3$ and NaOH solutions to form geopolymer paste that binds the aggregates and other un-reacted materials. The prepared geopolymers were characterized using combination of physico-mechanical and structural analyses. The results of cured sample after 7 and 28 days show equivalent compressive strength as compared to the conventional PC concrete.

Keywords: Geopolymer, alkali activator, AOD slag, redmud, flyash

*Corresponding author: e-mail: drpkr.chem@gmail.com

## Introduction

"Geopolymer" was coined by Davidovits in 1978 to represent a broad range of materials characterized by chains or networks of inorganic molecules (Geopolymer Institute 2010) [1]. There are nine different classes of geopolymers, but the classes of greatest potential application for transportation infrastructure comprised aluminosilicate materials that may be used to completely replace portland cement in concrete construction (Davidovits 2008) [2]. These geopolymers rely on thermally activated natural materials (e.g., kaolinite clay) or industrial byproducts (e.g., fly ash or slag) to provide a source of silicon (Si) and aluminum (Al), which is dissolved in an alkaline activating solution and subsequently polymerizes into molecular chains and networks to create the hardened binder. Such systems are often referred to as alkali-activated cements or inorganic polymer cements.

Geopolymer concrete (GPC), an innovative material characterized by long chains or networks of inorganic molecules, is a potential alternative to conventional portland cement (PC) concrete for use in transportation infrastructure construction. It relies on minimally processed natural materials or industrial byproducts to significantly reduce its carbon footprint, while being very resistant to many of the durability issues that can plague conventional concrete. However, the development of this material is still in its infancy. These inorganic products consist of the $[SiO_4]^{-4}$ and $[AlO_4]^{-4}$ tetrahedral networks. The influences of alumino-silicate materials, such as metakaolin and low calcium fly ash, were investigated by many researchers [3,4,5,6,7,8]. Their properties are suitable for reformed aggregate as ordinary Portland cement (OPC). Based on the past results, high calcium fly ash has been successfully used as a raw material in the geopolymer mixture [9,10]. Wide range of applications of Geopolymers [11,12,13,14,15,16] in the fields of new ceramics, binders, matrices for hazardous waste stabilization, fire-resistant materials and high-tech materials is a emerging trend of current research.

With the increased use of cement in concrete, there have been environmental concerns both in terms of damage caused by the emission of carbon dioxide during cement manufacture. About 1.5 tonnes of raw materials is needed in the production of every tonne of Portland cement, at the same time about one tonne of carbon dioxide ($CO_2$) is released into the environment during the production [17]. The production of PC is extremely resource and energy intensive process. Additionally, the global warming also can occur because of the greenhouse gases such as carbon dioxide emission to the atmosphere. Several studies have been carried out to reduce the use of PC in concrete to address the global warming issues. This has brought pressures to reduce the cement consumption in the construction industry. An attempt in this regard is the development of geopolymer concrete (GPC), which seems to be the future for construction industries in near future. Geopolymer is an innovative binder and can be produced by synthesizing from materials of geological origin or from by-product, which are rich in silicon and aluminium with highly alkaline solution [18].These include the utilization of supplementary cementing materials such as fly ash, redmud, blast furnace slag, rice-husk ash and metakaolin, are alternatives to Portland cement [19].

Aim of present investigation is to explore the possibilities of preparing geopolymers by using waste materials such as calcium bearing slag, flyash and redmud. The mixed materials are activated with alkali liquids essential for geopolymerization process. The activated process is accelerated by low temperature heat treatment slightly above the room temperature. The Geopolymers formed are cast in mold for properties evaluation at certain interval of time.

## Materials and Method
### Materials

Three waste materials are used for the preparation of geopolymers. They are calcium bearing slag, flyash, redmud. Calcium bearing slag is obtained from BRG steel plant (Odisha, India), which manufactures stainless steel by AOD (Argon oxygen decarburization) process. Flyash and redmud are obtained from Aluminium Manufacturer, Nalco (National Aluminium Co.) located at Bhubaneswar, India. Other raw materials used are sodium silicate ($Na_2O : SiO_2 > 2$-$2.2$), Sodium hydroxide (AR grade, purchased from Mumbai, India) washed clean sand water. Cement was purchased (ACC Limited) from local market, to prepare the cylindrical block to compare the properties with Geopolymer. Chemical composition of raw materials are shown in Table I.

**Table I.** Chemical composition of AOD slag and Redmud

| Raw Material | $SiO_2$ | $Al_2O_3$ | CaO | MgO | FeO | $Fe_2O_3$ | $TiO_2$ | $Cr_2O_3$ | $Na_2O$ | MnO | $K_2O$ | LOI | Basicity |
|---|---|---|---|---|---|---|---|---|---|---|---|---|---|
| AOD Slag | 30.5 | 5.88 | 45.86 | 10.78 | 0.68 | - | - | 1.5 | - | 3.5 | - | - | 1.95 |
| Redmud | 5.7 | 19.8 | 1.2 | Trace | - | 51.1 | 4.1 | - | 4.6 | - | - | - | 13.5 |
| Flyash | 62 | 27 | 2 | 1.0 | 0.5 | 0.5 | - | - | Trace | 0.5 | - | - | - |
| Sand | 83.06 | 6.01 | 0.45 | - | 4.5 | - | - | - | 0.33 | - | 3.2 | 1.09 | - |

**Chemistry of the resource materials**

XRD and SEM with EDX analyses of raw materials are shown in Figure 1 & 2.

Chemical analysis (Table 1) shows, AOD slag is rich in CaO (45.86%) and $SiO_2$ (30.5%) lower amount of $Al_2O_3$ (5.88%). The XRD analysis presence of Dicalcium silicate (C2S), Tricalcium silicate (C3S), Tricalcium aluminate (C3A) phases are also observed in AOD slag (Figure 1A) which provides opportunity for hardening as in case for commercial Portland cement, which has ample amount of C2S phases. Polymerization with similar molecular structure in AOD slag may aid breaking of silicate molecules to single mers and enhances the process and geopolymerization formation.

Redmud has been chosen as one of the additives for the preparation of Geopolymer for the following reason;
(i) It contains predominantly $Fe_2O_3$ (major), Alumina (19-20%), Silica (7-9%) and $TiO_2$ as evident from X-ray diffraction patterns (Figure 1C)
(ii) It is alkaline in nature $Na_2O$ (08-12%)
(iii) Calcium oxide is low (1-2%)

SEM study at different magnifications with EDX show the presence of Fe, Ti and Al. Looking at its composition and availability as a waste becomes an useful addition to AOD slag.

Indian flyash employed in the present investigation has high silica (62%) and moderately higher alumina content (27%). This chemistry creates condition favorable for development of Geopolymer with comparatively higher strength. The XRD of flyash reveals mostly the oxide phases such as silica and alumina. More interestingly, it also contains mullite as one of the phase contributing to the condition necessary for formation of Geopolymer.

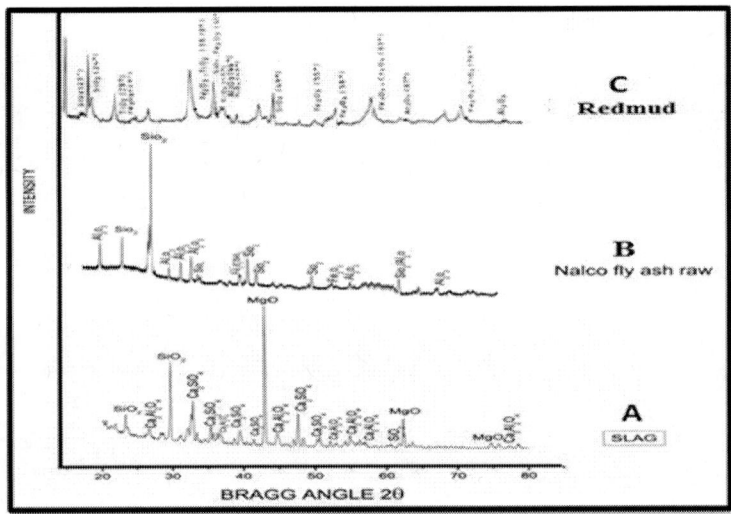

**Figure 1.** XRD of (A) AOD Slag, (B) Flyash and (C) Redmud

## Experimental

Geopolymer pastes are prepared by mixing slag with redmud and flyash individually with the alkali solution (a solution of 10% Sodium silicate solution and 10M NaOH) in certain ratio (Table II). The geopolymer paste is divided into two parts, one part for measuring the setting time and the other part of the mix is cast on ramming machine as per the standard size. The casted samples are kept in hot air oven for 24 hours at 70 °C and then in ambient atmosphere. The flow chart of the geopolymer preparation is shown in the Figure 3.

After the gap of 7 days and 28 days, samples are tested for compressive strength. The selected test pieces are examined by scanning electron microscope (SEM) with energy dispersive X-ray (EDX) analysis.

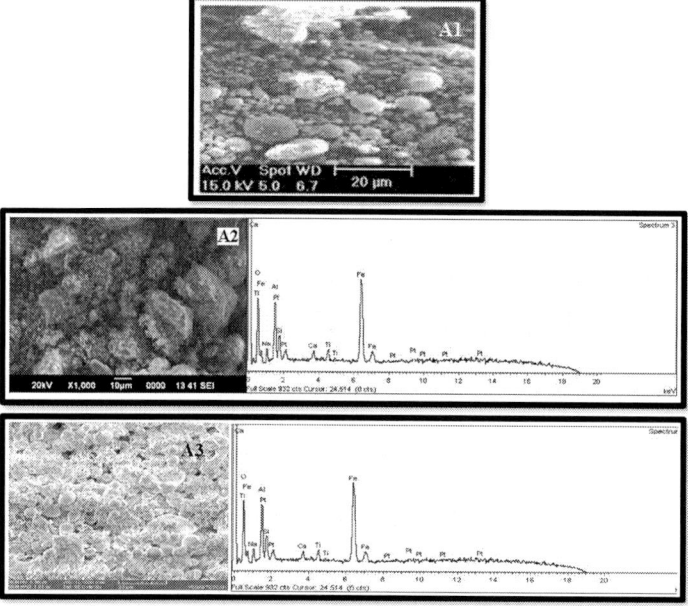

**Figure 2.** SEM/EDX figures of (A1) AOD Slag, (A2) redmud and (A3) Fly ash

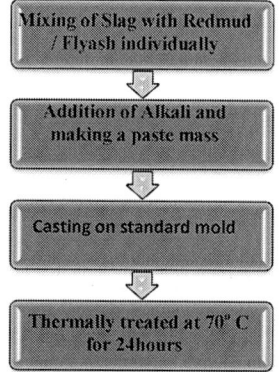

**Figure 3.** Flow chart for producing Geopolymer

Table II: Mix proportions of Geopolymers

| Sample Code * | AOD Slag (Wt.%) | Redmud (Wt.%) | Flyash (wt, %) | Portland Cement (wt, %) | Fine Sand (Wt.%) | Alkali (10%, Sod. Silicate + 5%, 10 M |
|---|---|---|---|---|---|---|
| | Avg. particle size of Ingredients | | | | | |
| | 210 μm | 149 μm | 149 μm | | 45 μm | |
| S1 | 30 | 25 | 0 | | 30 | 15 |
| S2 | 35 | 20 | 0 | | 30 | 15 |
| S3 | 40 | 15 | 0 | | 30 | 15 |
| S4 | 45 | 10 | 0 | | 30 | 15 |
| S5 | 50 | 05 | 0 | | 30 | 15 |
| S6 | 55 | 00 | 0 | | 30 | 15 |
| S7 | 27.5 | 27.5 | 0 | | 30 | 15 |
| PC | - | - | - | 55 | 30 | 15 % (Water) |
| S8 | 30 | 0 | 25 | | 30 | 15 |
| S9 | 35 | 0 | 20 | | 30 | 15 |
| S10 | 40 | 0 | 15 | | 30 | 15 |
| S11 | 45 | 0 | 10 | | 30 | 15 |
| S12 | 50 | 0 | 05 | | 30 | 15 |
| S13 | 27.5 | 0 | 27.5 | | 30 | 15 |

- S1 to S7 : Slag-redmud based Geopolymers and S8 to S13 Slag-Flyash based Geo-polymers

**Setting time of GP Paste**

Setting time of geopolymer was tested as per ASTM C 191-08. The paste was taken as prepared earlier in a bowl and cast into the conical mold of Vicat apparatus. Penetration of Vicat needle (1.00 ± 0.05 mm in diameter) in the GP paste was measured at regular intervals. The time for 25mm penetration was determined by interpolation, which represents the initial setting time. The final setting time of the GP paste was also recorded when the needle left negligible mark on the paste surface. The test was done at a temperature of room temperature around 28-30 oC.

**Compressive strength**

In this study, compressive strength test was performed on 70x50 mm cylindrical specimens using 2000 KN Digital Compressive Testing Machine (Mechatronics Control System; Model : MECH-CS/UTE 20T). A set of three specimens for each sample were tested at the age of 7 and 28 days.

**Scanning electron microscopy (SEM) with Energy dispersive X-ray (EDX)**

Scanning electron microscopy (SEM) with Energy dispersive X-ray (EDX) structural investigations were conducted using JEOL, Japan, Model: JSM-6390 LV, in order to get a better understanding of the morphology of the geopolymer as well as to determine the effect of redmud ratio into slag on the properties and microstructure.

## Results and Discussions

The effect of setting time on the development of compressive strength and surface morphology were studied at ambient temperature as follows.

### Setting time

The setting time for redmud based and flyash based geopolymers are shown in figure 4. The initial setting time is high for sample S6. The setting time has gradually decreased with the addition of redmud and flyash singly into slag. It is inferred that the higher the slag content in the paste, slower is the setting process which is due to presence of high calcium oxide in the slag. Among slag-redmud and slag-flyash combinations, the slag based flyash geopolymer has shown less setting time as compared to slag-redmud based GP, which may be due to higher Alumina in Geopolymer. The fine slag particles having high surface area react faster and make the setting time shorter. This finding is in agreement with the finding of Chindaprasirt P et.al [20].

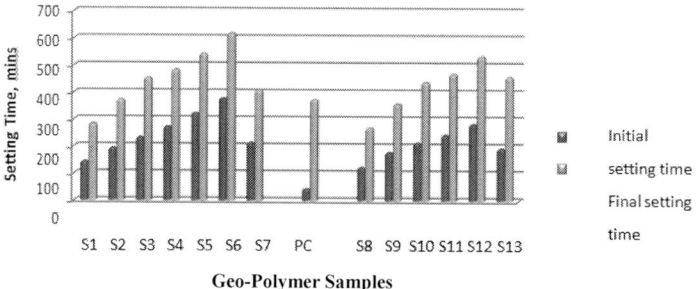

Figure 4. Setting time of redmud and flyash based Geopolymers

### Compressive strength of Geopolymers

The compressive strength was tested for each geopolymer samples at different time intervals. At least three specimens were tested for one sample at an age of 7 days after casting and also 28 days. The finding was very interesting, (figure 5). An abnormal difference was notriced in compressive strength values for specimens cured at $7^{th}$ day and $28^{th}$ days for both at slag-redmud and slag-flyash based geopolymers. At the beginning after 7 days of curing, there was an increasing trend for compressive strength of Geopolymers from 4.8 to 8.2 MPa on addition of redmud to the mixture and the same trend was also observed in case of 28 days i.e from 28.5 to 45.6 MPa. But for slag-flyash based samples there was a increase in compressive strength i.e., to 52.7 MPa as compared to slag-redmud samples. The

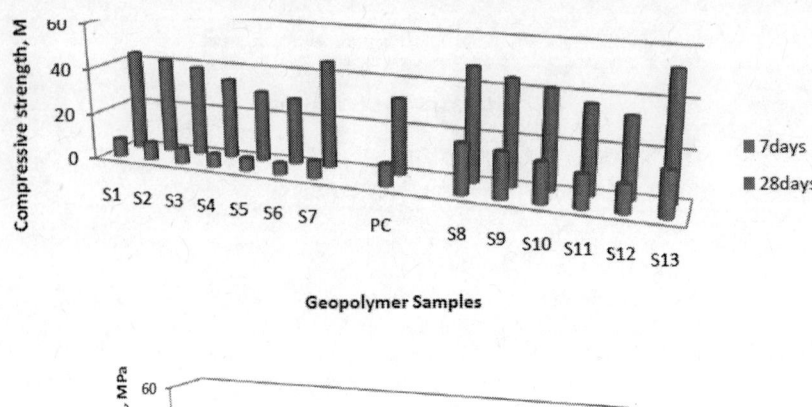

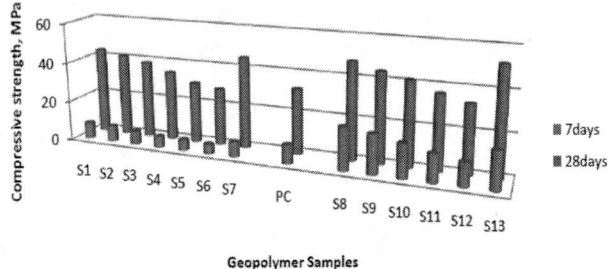

Figure 5.   Compressive strength of Geopolymers

**Microstructure of Geopolymers**

The SEM/EDX results in Figure 6 indicate that the major elements are Silica, Alumina and Calcium oxide and Sodium oxide in geopolymers. The presence of Na and Ca confirmes that slag based geopolymer have these ions in amorphous region, which may be the reason for slow setting of Geopolymer. Figures 2 (A1), (A2) and (A3) show the major elements present in the raw materials such as AOD slag, redmud and flyash, which play vital role in the polymerization process. The gel structures formed during alkali activation are partially glassy phase which can be seen in Figures 5 (C), (D) and (E).

The EDX data shows atomic ratio of Si/Al in the matrix which are 4.06, 4.52 and 3.16 for samples S1, S5 and S7 respectively. From these analyses it can be said that increase in Si/Al ratio is accompanied by decrease of compressive strength of the geopolymers. There is strinking similarity between the present finding with the finding reported by T. Bakharev et.al. [22]. The cracks formed around the borders of non-dissolved particles and in the gel phase surrounding the particles (Figures 6 C & E) indicate that the mechanical strength of the geopolymeric binder was lower than that of the non-dissolved particles. Generally, the interface between the nondissolved particles and the gel phase is the most sensitive for the materials failure during the compressive strength tests. The compressive strength is found to be higher for both slag-redmud and slag- flyash samples (S7 and S13), i.e. 45.5 MPa and 52.7 MPa, due to appreciable ratio of Si/Al in the matrix.

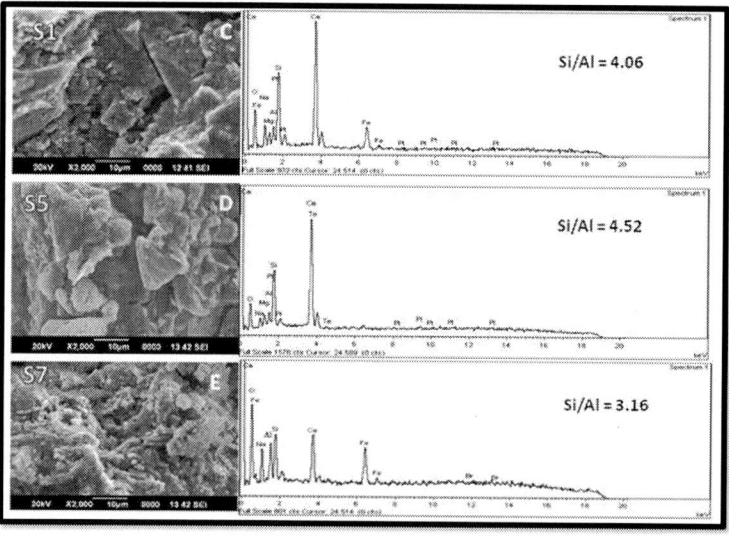

**Figure 6:** SEM/EDX figures of (C) S1, (D) S5 and (E) S7

## Conclusions

In the present work, the geopolymerization of high Ca content AOD slag produced under Indian condition and red mud and flyash are investigated. The concentrations of AOD slag to redmud or flyash in the aqueous phase of the system are proved as crucial synthesis parameters. Under the optimum synthesis conditions defined in this work, the AOD slag-redmud and AOD slag-flyash based inorganic polymeric materials has developed compressive strength, 45.6 MPa and 52.7 MPa respectively after 28 days of curing. Based on these test results, the following conclusions are drawn:

1. AOD slag based geopolymers cured in the ambient temperature after laboratory oven setting at 70 °C, gains compressive strength at par with PC. The 28th day compressive strength of ambient-cured specimens shows nearly equivalent compressive strength with PC.
2. The decrease in setting time in the geopolymer may be due to excess amount of Ca content in the AOD slag.
3. Due to the high compressive strength, this material (Geopolymer) can be used in Aerospace and surface engineering applications.

## Acknowledgement

We are grateful to the Chairman cum Managing Director of BRG Steels Pvt. Ltd. for supplying the AOD slag for the experimental work. We are also grateful to Chairman and Secretary of Gandhi Institute of Engineering and Technology, Gunupur, Odisha, for their support in this research.

## References

[1] *What Is a Geopolymer? Introduction* (Geopolymer Institute, Saint-Quentin, France. Accessed on January 29, 2010) http://www.geopolymer.org/science/introduction.

[2] Davidovits. J, *Geopolymer Chemistry and Applications* (Geopolymer Institute, Saint- Quentin, France, 2008)

[3] Xu H and Van Deventer JSJ, *The geopolymerisation of natural alumino-silicates, Proceedings: 2nd International conference on geopolymere'99* (Geopolymer Institute, 1999), 43-63

[4] Palomo A, Grutzeck MW and Blanco-Varela MT, *Alkali activated fly ashes: cement for the future, Cement and Concrete Research* ( Elsevier, 1999) 1323-1329.

[5] Xu H and Van Deventer JSJ, *Geopolymerisation of multiple minerals* (Minerals Engineering, Elsevier, 2002) 1131-1139.

[6] Fernandez-Jimenez. AM, Lachowski. Palomo EE. A and Macphee. DE, "Microstructural characterization of alkali-activated PFA matrices for waste immobilization", *Cement and Concrete Composites, Elsevier* (2004), 1001-1006.

[7] Bakharev. T, "Geopolymeric materials prepared using Class F fly ash and elevated temperature curing", *Cement and Concrete Research, Elsevier* (2005), 1224-1232.

[8] Andini. S, Cioffi. R, Colangelo. F, Grieco. T, Montagnaro. F and Santoro. L "Coal fly ash as raw material for the manufacture of geopolymer-based products", *Waste Management, Elsevier* (2008), 416-423.

[9] Chindaprasirt. P, Chareerat. T, and Sirivivatnanon. V "Workability and strength of coarse high calcium fly ash geopolymer", *Cement and Concrete Composites, Elsevier* (2007), 224-229.

[10] Sathonsaowaphak. A, Chindaprasirt. P and Pimraksa. K "Workability and strength of lignite bottom ash geopolymer mortar", *Journal of Hazardous Materials, Elsevier* (2009), 44-50.

[11] Van Jaarsveld. JGS, van Deventer. JSJ, Lorenzen. L "The potential use of geopolymeric materials to immobilise toxic metals" *Part I. Theory and applications. Miner. Eng.* (1997), 10, 659–669.

[12] Kriven. WM, Bell. JL and Gordon M "Geopolymer refractories for the glass manufacturing Industry" *Ceram. Eng. Sci. Proc.* (2004) 25, 57–79.

[13] Lancellotti, I, Kamseu. E, Michelazzi, M, Barbieri. L, Corradi and A, Leonelli. C "Chemical stability of geopolymers containing municipal solid waste incinerator fly ash" *Waste Manag.* (2010) 30, 673–679.

[14] Andini. S, Cioffi. R, Colangelo. F, Ferone. C, Montagnaro. F and Santoro, L. "Characterization of geopolymer materials containing MSWI fly ash and coal fly ash". *Adv. Sci. Technol.* (2010) 69, 123–128.

[15] Giancaspro. JW, Papakonstantinou. CG, Balaguru. P "Flexural response of inorganic hybrid composites with E-glass and carbon fibers" *J. Eng. Mater. Technol.*(2010) *132*, 1–8.

[16] Ferone. C, Colangelo. F, Cioffi. R, Montagnaro. F and Santoro. L "Mechanical performances of weathered coal fly ash based geopolymer bricks" *Procedia Eng.*(2011) 21, 745– 752.

[17] McCaffery. R, *Climate Change and the Cement Industry*, Global Cement and Lime Magazine, Environment Special Issue (2002), 15-19.

[18] Davidovits. J, *Chemistry of geopolymeric systems, terminology* (Proceeding of Geopolymer '99 International Conference, Saint-Quentin, France, 1999).

[19] Rangan, BV. *Studies on Fly Ash-Based Geopolymer Concrete*, Malaysian Construction Reasearch Journal, Vol. 3 (2008).

[20] Chindaprasirt. P, Homwuttiwong. S, and Sirivivatnanon. V "Influence of fly ash fineness on strength, drying shrinkage and sulfate resistance of blended cement mortar." Cem. Concr. Res., 34(7) (2004), 1087–1092.

[21] Kamhangrittirong. P, Suwanvitaya. P and Suwanvitaya. P "Synthesis and properties of High Calcium fly ash based geopolymer for concrete applications", Chindaprasirt, 36th Conference on Our World in Concrete & Structures, Singapore, August 14-16, 2011.

[22] Bakharev. T, Geopolymeric materials prepared using Class F fly ash and elevated temperature curing, Cement and concrete research, Elsevier (2005), 1224-1232.

# SYNTHESIS OF COMPOSITE TaC-TaB$_2$ POWDERS

Behzad Mehdikhani[*], Gholam hossein Borhani, Saeed Reza Bakhshi

*Department of Materials Engineering,*
*Malek-e-ashtar University of Technology,*
Isfahan, **Iran**

Hamid Reza Baharvandi

*Department of Materials Engineering,*
*Malek-e-ashtar University of Technology,*
Tehran, **Iran**

## Abstract

In this study, mechanochemical process (MCP) is applied to synthesize ultrafine TaC powders. In this research, nano composite powder TaC-TaB$_2$ was produced using mixtures of tantalum carbide and boron carbide as raw materials via mechanochemical process. The phase formation characterization during process was utilized by X-ray diffractometer (XRD). The morphology of synthesized powder was studied using scanning electron microscopy (SEM).

**Key words:** Mechanochemical process (MCP), Tantalum carbide, TaC-TaB$_2$ composite.

## Introduction

Development of the ceramic matrix composites (CMCs) that combine advantageous properties of the individual components for high performance applications is of special interest to the ceramic industry [1]. For example, the composites of boride and carbide offer anattractive combination of excellent mechanical and electrical properties,which renders them promising candidates for the advancedstructural applications at high temperatures and/or in corrosiveenvironments.With the rapid development of aerospace technology, it is urgent to study the new material used in high temperature structural components[2].Tantalum Carbide (TaC) is an ultra high-temperature ceramic (UHTCs) for high performance applications with a melting point in excess of 3900°C[1-3]. Unique combination of good chemical stability, good corrosion resistance, high modulus (537GPa), high hardness(15–19GPa) and other good mechanical properties make it a candidate material for rocket propulsion components [4-5] and many applications as high speed cutting tools, wear resistant parts, hard coating on hard metals and high temperature structural materials[1]. Performance of this material at high temperature needs good oxidation resistance [5]. Researchers have been found that $TaB_2$ is suitable additives [4]. Tantalum borides have not been studied as extensively as other borides, like $TiB_2$ [6], $ZrB_2$ [7], and $HfB_2$ [8].According to the Ta–B phase diagram [9], there are five boride phases including $Ta_2B$, $Ta_3B_2$, $TaB$, $Ta_3B_4$, and $TaB_2$. Besides, a new phase $Ta_5B_6$ and its crystal structure were later identified by Bolmgren et al. [10]. On the formation of tantalum borides, a variety of processing routes have been utilized. For example, Hideaki et al. [11] produced $TaB_2$, $TaB$, and $Ta_3B_4$ by solid state reactions of mixed Ta and amorphous boron powders under corresponding compositions at 800, 900, and 1800°C, respectively. Peshev[12] obtained $TaB_2$ through borothermic reduction of$Ta_2O_5$ at 1650°C for 1 h. The tantalum diboride ($TaB_2$) with comparable mechanical properties to $ZrB_2$ and $HfB_2$ was also fabricated by reducing $Ta_2O_5$ with $B_4C$ and graphite at 1600°C.The ternary phase diagram is known since the 1963 in Fig. 1[13]. It is well known that the properties of composite $TaC-TaB_2$ are strongly dependent on the relative amounts and types of the various phases formed. It has to be noted that $B_4C$, which is also a component of the Ta–C–B ternary system, is not chemically compatible with Ta or TaC. It reacts with Ta and TaC forming $TaB_2$and free carbon that has been used for the synthesis and processing of diboride-containing ceramics [14]. Recently, mechanical activation and mechanical milling have been extensively used for synthesis of advanced materials. The technological advantages of the mechanochemical synthesis are obvious. High-temperature processes and furnaces for synthesis of this composite are avoided. The product obtained is finely dispersed and has a defect structure which increases the sinterability of $TaC-TaB_2$ powder. This article demonstrates the formation of $TaC-TaB_2$ by ball milling a mixture of TaC and $B_4C$ with the stoichiometry of reaction (1):

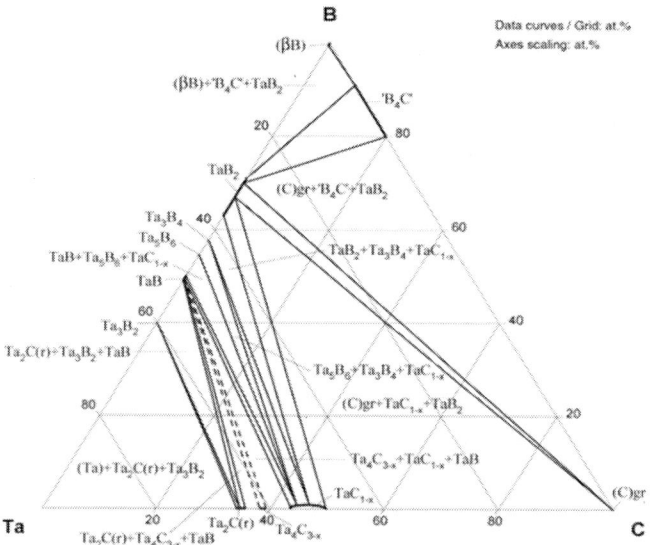

Figure 1. The ternary diagram B-Ta-C [14].

TaC + B$_4$C = TaB$_2$+3C    $\Delta H°_{298}$ = −59.67 kJ/mol Eq.(1)
The negative value of $\Delta H°_{298}$ suggests that this reaction is exothermic and should be self-sustaining.

## Experimental

### Raw materials

The starting powders were TaC (Ningxia Orient Tantalum Industry Co., Ltd. China), and B$_4$C (Jingangzuan Boron Carbide Co. Ltd., China) with mean particle size1.25μm and300 nm, respectively. The TaC content in the TaC powder was higher than 99%. The main impurities were ~0.3 wt% Nb, 0.1 wt% Fe, 0.20 wt% O,0.15 wt% free carbon, 0.05 wt% N, and Al, Ca, K, Na, Ti with a total amount <0.05 wt%. The B$_4$C was >95 wt% pure with major impurities of free carbon.

### Powder preparation

25g TaC and 2.0%Wt. B$_4$C were mixed in a WC cylinder (250ml) using 150grWC balls as mixing media. Ball milling of powder mixture was carried out in a planetary ball mill at room temperature in n-hexane and under argon atmosphere. The ball-to-powder weight ratio and the rotational speed of vial were 20:1 and 250 rpm, respectively for 3, 6 and 12 hours.The milling was interrupted at selected times and a small amount of powder wasremoved for further characterizations.

## XRD analysis

Phase transformation during milling were determined by X-ray diffraction (XRD) in a Philips X'PERT MPD diffractometer using filtered Co K$\alpha$ radiation ($\lambda$ = 0.178 nm). The lattice parameter of a cubic substance is directly proportional to the spacing 'd' of any particular set of lattice planes (eq: 2) [15].

$$a = \frac{d}{\sqrt{h^2 + k^2 + l^2}} \qquad \text{Eq. (2)}$$

the Nelson–Riley method was used to minimize errors caused by aberration of 2$\theta$ variation and the lattice parameter 'a' of TaC was calculated for at least three peaks, using Eq. (3) [16].

$$F(\theta) = \frac{1}{2}\left(\frac{\cos^2\theta}{\sin\theta} + \frac{\cos^2\theta}{\theta}\right) \qquad \text{Eq. (3)}$$

Silicon standard sample with large grains and free from defect broadening was used as a standard to increase the precision of the instrumental broadening. Then, the error of diffractometer was eliminated by Eq. (4) [27].

$$b = b_{size} + b_{strain} = \sqrt{b_0^2 - b_s^2} \qquad \text{Eq. (4)}$$

Where '$b_s$' is the FWHM of the main peak of Si standard sample (2theta=28.5°) used for calibration and '$b_0$' is the FWHM of TaC's peaks. Both '$b_s$' and '$b_0$' were calculated by X-Pert High Score software. Crystallite size and internal strain of specimens were calculated from broadening of XRD peaks using the Williamson-Hall method [17].

$$B \cos\theta = K\lambda/d + \mu \sin\theta \qquad \text{Eq. (5)}$$

Where $\theta$ is the Bragg diffraction angle, d is the average crystallite size, k is a constant (with a value of 0.9), $\lambda$ is the wavelength of the radiation used, and $\beta$ is the diffraction peak width at half maximum intensity. The morphology and microstructure of milled powder particles were examined by SEM images in a Philips XL30 at an accelerating voltage of 30 kV.

## Morphology of powders

The morphology of selected mechanically alloyed powders was examined by a Scanning Electron Microscope (SEM-Philips XL30) operating at 30 kV.

## Results and Discussion

### X-ray Diffraction Analysis

A commercial software program (HSC Chemistry) was used to identify the probable reaction using thermodynamic data.Fig. 2 show X-ray diffraction patterns of the TaC and TaC-TaB$_2$ powder after various milling time. Figure (2-a) shows XRD pattern of raw tantalum carbide powder without milling. Figure (2-b).shows XRD pattern of powder after 3 hours milling were identified as a mixture of starting materials.The results show that the powder milled for 3 hours, no new phase is not formed.B$_4$C phase in the XRD spectra have been observed in some detail because it is a small amount of 2% by weight. Figure (2-c) shows the XRD results of the powders milled for 6 hours, TaB$_2$phase unformed and there are still some B$_4$C composition.The XRD analysis indicated that only TaB and TaB$_2$, as presented in Figure (2-d), were produced from the samples of their equivalent stoichiometry's. Figure (2-d) shows that the final products obtained from mechanochemical synthesis was free from three Ta-rich samples of Ta$_2$B= 2:1, Ta$_3$B$_2$=3:2 and Ta$_3$B$_4$=3:4 contain a large amount of residual Ta. No XRD peaks were observed for B$_4$C, suggesting that a reaction had occurred in the system. Cup and balls of ball mille was made from WC so X-ray results show that increasing milling time up to 12 hours has caused a spike tungsten carbide. Figure (2-d).

The lattice parameter of TaC powder obtained from 3, 6 and 12 h milling is calculated by Nelson–Riley method from the XRD analysis (Fig.3). In this method, the accurate lattice parameter is obtained by extending the lattice parameter function and its connection to zero content of Nelson–Riley parameter. Fig.3 shows that the lattice parameter increases by increasing milling time.

The change of lattice parameter of milled samples for 3, 6 and 12 h is presented in Table 1. From Table 1, it is concluded that there is little decrease in the lattice parameter by increasing milling time.

Table 1. The lattice parameter of TaC powder after 3, 6 and 12 h milling times.

| Milling time, h | Lattice parameter, Å Nelson–Riley method |
|---|---|
| 3 | 4.3242 |
| 6 | 4.4063 |
| 12 | 4.4526 |

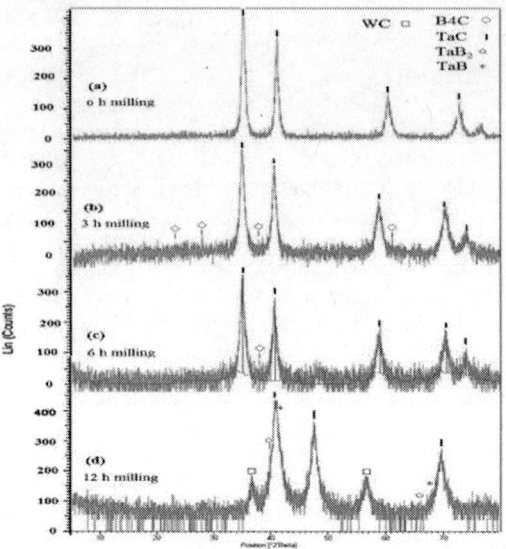

**Figure. 2.** XRD patterns of TaC-B4C powder after various milling time:
(a) 0, (b) 3, (c) 6, and (d) 12 h.

The grain size 'd' and strain 'µ' of TaC products during milling were measured by the Williamson–Hall method (Figure 4). The FWHM of the diffraction peak is wider with milling time due to the strain and the refinement of powder.A plot of B cos θ versus sin θ [18] is shown in Fig.4. The average grain sizes of the TaC calculated from the XRD data were about 801, 267, 94 and 34 nm for the samples with milling times of 0, 3, 6, and 12 h. According Eq. (5) and Williamson-Hall method calculation, internal strain and crystallite size of milled powder is shown in Fig.5. Internal strain of powders has increased with increasing milling time and crystallite size is reduced.

## Microscopic examinations

SEM images of TaC powder with milling time are shown in Fig. 6.This figure shows morphology of powder particles after different milling times. SEM images of TaC powder with milling time are shown in Figure 6. The TaC powder without milling has an angular shape. Figure (6-b) shows that particle size reduction after 3 h of milling. At the beginning of the milling process, the brittle particles (TaC) got fragmented and comminuted and TaC powder gets rounder shape and comes refinement with milling time. Figure (6-c) shows that after 6 h milling the powder particles became nearly equiaxed in shape with a wide size distribution of 0.3-1µm.By increasing milling time to 12 h the rate of fracturing increased and as a result the size of powder particles decreased. At this stage, the morphology of powder particles was almost equiaxed Figure (6-d). The powder particles after 12 h of milling time are large agglomerates of ultrafine particles ranging from 0.1 µm to 2 µm.

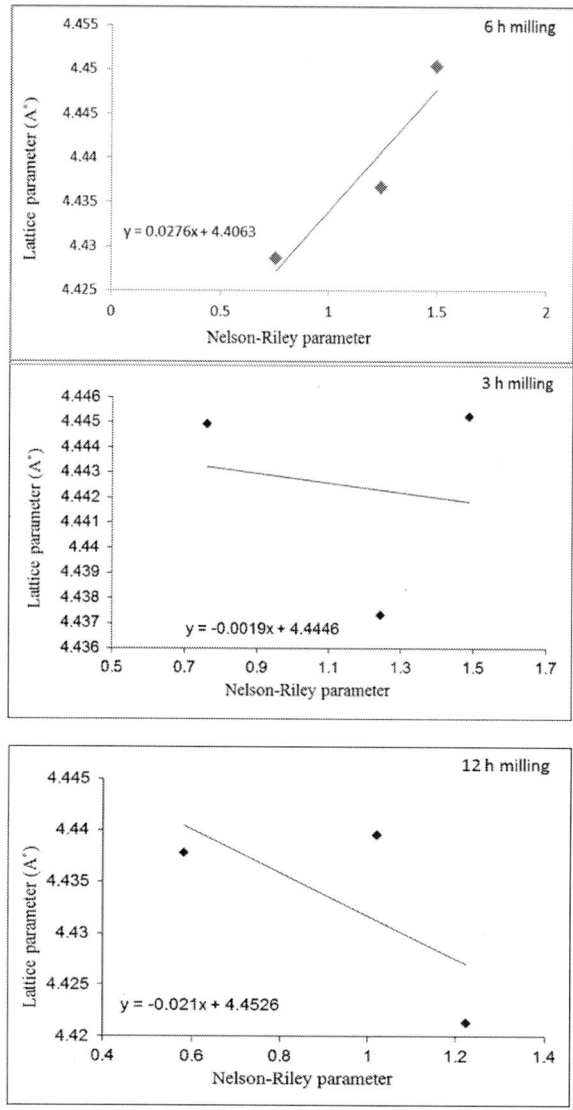

**Figure 3.** Nelson–Riley parameter for obtaining the lattice parameter of TaC phase after 3, 6 and 12 milling times.

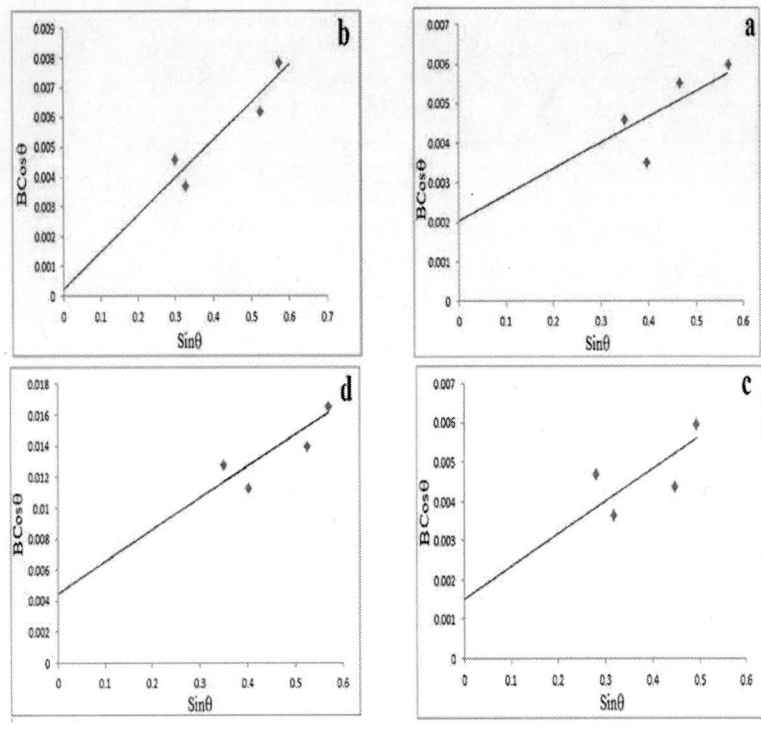

**Figure.4.** Plot of B cos θ versus sin θ for TaC-B$_4$C milled powders:
(a) 0, (b) 3, (c) 6, and (d) 12 h.

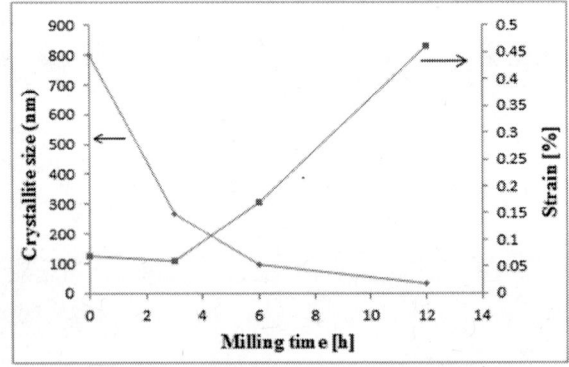

**Figure 5.** Effect of milling time on the crystallite size and internal strain of powders.

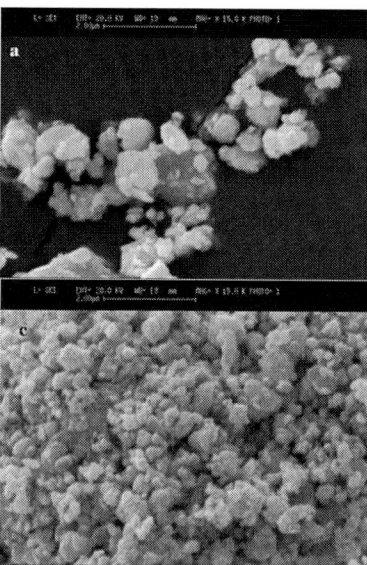

Figure.6. Scanning electron micrographs of starting powders particles
(a) TaC powder,
(b) after 3 h,
(c) after 6 h, and
(d) 12 h of milling times.

## Conclusions

TaC-TaB$_2$ composite powder has been synthesized by a low temperature solid-state reaction between tantalum carbide and boron carbide. Studies show that by milling of powders TaC and B$_4$C, TaB$_2$ phase and the resulting composite TaC-TaB$_2$ is composed. Prolonging the milling time up to 12 h resulted in a decrease of grain size to nano scale along with increasing strain and a slight increase in lattice parameter of TaC phase. Crystallite size of milled powder decreased with increasing milling time up to 3, 6 and 12 h. the powder particle size reached to 267 were 94 and 34 nm. TaB$_2$ phase in the powder milled for 12 h was observed, indicating that some mechanochemical synthesis process is carried out. Increasing milling time was increased internal strain of crystallites.

## Acknowledgements

The authors are indebted to material department in Malek-e-ashtar University of Technology that supplied the raw materials for the development of this research and to building and construction department of Standard Research Institute for its equipment support.

## References

1. T.H. Squire and J. Marschall, *Journal European Ceramic Society*. 30, 2239-2251(2010).

2. F. Monteverde, A. Bellosi and L. Scatteia, *Materials Science and Engineering*. 485, 415–421 (2008).

3. D. Hwan Kwon, S. H. Hong, B. K. Kim, *Journal of Materials Letters*. 58, 3863– 3867 (2004).

4. X. Zhang, G. E. Hilmas,W. G. Fahrenholtz, *Materials Science and Engineering*. 501, 37–43 (2009).

5. G. Hagemann, H. Immich, T. V. Nguyen, E. Gennady, p 14 (1998).

6. R.G. Munro, J. Res. National. Institute of Standards Technology. 105, 709–720 (2000).

7. W.G. Fahrenholtz, G.E. Hilmas, I.G. Talmy, J.A. Zaykoski, *Journal of American Ceramic Society*., 90, 1347–1364(2007).

8. R. Licheri, R. Orru`, C. Musa, A.M. Locci, G. Cao, *Journal of Alloys and Compounds*, 478, 572–578 (2010).

9. T.B. Massalski, H. Okamoto, P.R. Subramanian, L. Kacprzak (Editors), ASM International, Materials Park, OH, USA, 1996.

10. H. Bolmgren, T. Lundstrom, L.E. Tergenius, S. Okada, I. Higashi, *Journal of Less Common Metals*, 161, 341–345(1990).

11. I. Hideaki, S. Yusuke, K. Satoshi, N. Shigeharu, *Journal Ceramic Society Japan*, 98, 264–268 (1990).

12. P. Peshev, L. Leyarovska, G. Bliznakov, *Journal of Less Common Metals* 15, 259–267 (1968).

13. Peter Rogl, Springer 2009.

14. I.G. Talmy, J.A. Zaykoski, M.M. Opeka, *Journal of European Ceramic Society*. 30, 2253–2263(2010).

15. B.D. Cullity, Addison-Wesley, Inc., Massachusetts, MA, USA., 1956.

16. J.B. Nelson, D.P. Riley, Proceedings of the Physical Society 57, 160–177(1945).

17. G. K. Williamson, W. Hall, *Acta Metallurgica*, 1, 22–31(1953).

18. C. Suryanarayana, M. Grant Norton, Plenum Press, New York, 1998.

# CHARACTERISTICS OF COMPOSITE MICROSTRUCTURES AND PHASES

Advanced Composites for Aerospace, Marine, and Land Applications
Edited by: Tomoko Sano, T.S. Srivatsan, and Michael W. Peretti
TMS (The Minerals, Metals & Materials Society), 2014

# LASER DEPOSITED IN-SITU TiC REINFORCED NICKEL MATRIX COMPOSITES: MICROSTRUCTURE and TRIBOLOGICAL PROPERTIES

Tushar Borkar, Sundeep Gopagoni and Rajarshi Banerjee[*]

Department of Materials Science and Engineering
**University of North Texas**
Denton, TX 76203, USA

Junyeon Hwang

Institute of Advanced Composite Materials
**Korea Institute of Science and Technology**
Jeonbuk, 565- 905, Korea

Jaimie Tiley

Materials and Manufacturing Directorate
**Air Force Research Laboratory**
Dayton, OH 45433, USA

## Abstract

A new class of Ni-Ti-C based metal matrix composites has been developed using the laser engineered net shaping (LENS) process. These composites consist of an in situ formed and homogeneously distributed titanium carbide (TiC) phase reinforcing the nickel matrix. Additionally, by tailoring the Ti/C ratio in these composites, an additional graphitic phase can also be engineered into the microstructure. The microstructure of these composites has been characterized in detail in three dimensions, via serial-sectioning in a dual-beam focused ion beam instrument followed by appropriate 3D reconstructions, to determine the true morphology and spatial distribution of the TiC and graphite phases as well as the phase evolution sequence. These three-phase Ni-TiC-C composites exhibit excellent tribological properties, in terms of extremely low coefficient of friction while maintaining a relatively high hardness, and these will be discussed in the presentation.

**Keywords**: Laser deposition; Focused Ion Beam (FIB); Nickel-Titanium-Carbon composites; 3D reconstruction; Phase transformation.

## Introduction

Metal matrix composites (MMCs) exhibit greater strength in shear and compression by combining the metallic properties (ductility and toughness) with ceramic characteristics (high strength and modulus) [1-4]. MMCs exhibit attractive physical and mechanical properties, such as high specific modulus, fatigue strength, thermal stability, and wear resistance that make them suitable for automotive and aerospace industries and other structural applications [1-2]. MMCs are also used in electronic applications due to their low co-efficient of thermal expansion and higher thermal conductivity [1, 5-7]. The properties of MMCs are mainly governed by two key factors: size and volume fraction of reinforcements as well as nature of matrix reinforcement interface [2,8]. *In situ* MMCs exhibit thermodynamic stability, good interfacial bonding, and uniform fine particle distribution in the metal matrix, which leads to better mechanical properties compared to conventional *ex situ* MMCs [2,9]. Nickel and nickel base superalloys are widely used in automotive and aerospace applications (aircraft jet engines, land base turbines, and chemical-petrochemical plants) due to their excellent properties such as high resistance to
corrosion as well as fatigue and low thermal expansion [8]. Titanium carbide (TiC) has a very high hardness (2859-3200 HV), high melting point (3420 K), low density (4.93 $g/cm^2$), and high mechanical strength, but it is very brittle and cannot be used as a monolithic ceramic [8, 10-14]. Therefore, TiC reinforced nickel matrix composites are considered as a good candidate for high temperature refractory, abrasive, and structural applications [8, 10-14]. Using laser engineered net shaping (LENS™), novel monolithic composites based on Ni-Ti-C have been developed
(U.S. patent Application No. 13/769,787) that combine properties such as solid lubrication (e.g. graphite), high hardness (e.g. TiC), and high fracture toughness (e.g. nickel), for structural as well as surface engineering applications. These multifunctional, monolithic composites are needed in industrial applications, such as drilling components (wear band, stabilizer, drill collar, etc.), tunnel boring, and land base turbines.

Three dimensional (3D) microstructural characterization is very important to reveal many microstructural parameters, such as the number of features per unit volume, true size, shape, and morphology of microstructural features, and the connectivity between features, that is not revealed by simple two dimensional (2D) microstructural characterization [15-17]. Tomographic techniques based on the 3D atom probe and transmission electron microscopy (TEM), focus ion beam (FIB) serial sectioning, synchrotron-based x-ray diffraction and tomography, and sequential mechanical polishing or micromilling are techniques particularly suited for characterizing the local structure, chemistry, and/or crystallography of materials in 3D, from the atomic scale all way up to the millimeter size volumes [15]. Serial sectioning is one of the most common and powerful methodologies used to obtain 3D microstructural data. The dual-beam FIB has become a very powerful tool for such serial sectioning and 3D microstructural reconstruction of metals with multiple phases exhibiting complex morphologies and crystallographic orientations [15-16, 18-21]. Typically in FIB serial sectioning, to characterize particular microstructural features of interest, the general rule of thumb is that there should be a minimum of 10-20 sections per feature to accurately represent its shape and size [15]. The present paper mainly focuses on 3D microstructural characterization of novel laser deposited Ni- Ti-C composites. The salient features of the present study are to investigate the 3D morphology, size scale and distribution of TiC and graphitic precipitates in a nickel matrix and the implications on the sequence of phase evolution during solidification.

## Experimental Details

### Laser Engineered Net Shaping (LENS™)

The titanium carbide (TiC) reinforced nickel based composites used for 3D microstructural characterization were deposited using the laser engineered net shaping (LENS™) process from a feedstock consisting of a blend of elemental nickel, titanium and nickel-coated graphite powders. The processing details for these composites have been described in detail in previous papers [8] but a short summary is included here for completion. The powders used for depositing the Ni-Ti-C composites consisted of commercially pure near spherical Ni (40-150 µm), pure Ti (40-150 µm) and Ni coated graphite powders, which were premixed in a twin roller mixer with the nominal compositions 80Ni-10Ti-10C, 73Ni-7Ti-20C, and 77Ni-3Ti-20C (refer to Ni-10Ti-10C, Ni-7Ti-20C and Ni-3Ti-20C for brevity). These composites were laser deposited in a cylindrical geometry of diameter 10 mm and height 10 mm. The LENS™ deposited *in situ* Ni-Ti-C composites were characterized by scanning electron microscopy (SEM) in FEI-Quanta Nova-SEM. X-ray diffraction analysis of deposits was performed using (1.54 Cu Kα) line of Rigaku Ultima III X-ray Diffractometer. The micro-hardness was measured using a standard Vickers micro-hardness tester using 300g load. Sliding friction and wear testing was conducted with a Falex (Implant Sciences) ISC-200 pin-on-disk (POD) system at room temperature. The samples were openly exposed in lab air (~40% RH) during the tests. Tests were performed under a 1 N normal load with a 1.6 mm radius $Si_3N_4$ ball, which correspond to an initial maximum Hertzian contact stress ($P_{max}$) of ~1.2 GPa. The sliding speed was fixed at 50 mm/sec. The ratio of tangential to normal load is the friction coefficient. At least three POD tests were run out to a total sliding distance of 140 m (steady-state friction behavior). Area fraction calculation of TiC and graphite reinforcements was done using ImageJ software, and an average of 50 SEM images were reported in this paper. The details of FIB serial sectioning and 3D reconstruction are explained in the next section.

### Dual Beam FIB-SEM Serial Sectioning Methodology

This section will briefly describe the DB FIB SEM serial sectioning methodology employed to study the 3D microstructure characterization of Ni-Ti-C composites. A dual beam workstation (FEI Nova nanosem) equipped with FIB column employing a Gallium (Ga) liquid metal ion source, combined with a high-resolution field emission scanning electron microscope (FEG SEM) was used. Prior to the 3D serial sectioning, the stage was tilted to maintain the 52° angle between the FIB and SEM column as shown in Figure 1a, so that the sample surface was oriented perpendicular to the ion beam during cross sectioning. A layer of platinum (Pt) was deposited on the top of region of interest, which serves not only as protection but also helps in minimizing curtaining effects and provides a sharp cutting edge during serial sectioning [15, 20, 22]. In addition to this, Pt deposited layer is used as fiducial marking during 3D reconstruction for alignment of obtained 2D SEM images. The sample has been FIB micro machined into cantilevered beam configuration as shown in Figure 1b prior to the serial sectioning experiment. This sample geometry has 2 main advantages: first to minimize the redeposition of milled material onto the surface of interest, and second is to eliminate the possibility of shielding the detectors [15]. The serial sectioning procedure has been automated for FIB nova Nano SEM using FEI's auto slice and view software Automation software is essential for this serial sectioning experiment as it ensures consistent serial-section removal rate, and this also eliminates the need for human supervision and interaction, so that the experiment can run for long period of time and contain potentially large number of serial sections [15].

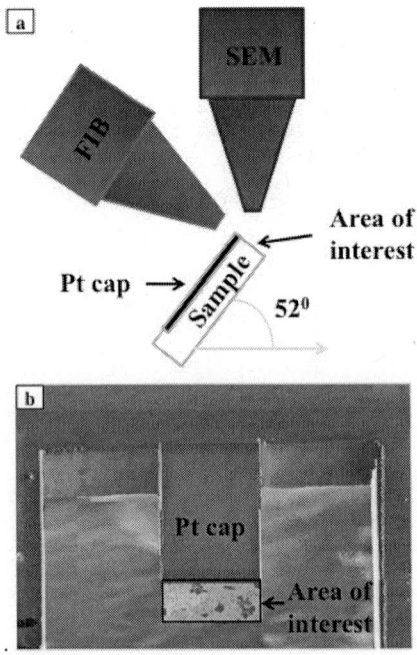

**Figure 1.** (a) Schematic of the sample geometry relative to the FIB and SEM columns for serial- sectioning
(b) Representative cantilever beam sample geometry used for serial -sectioning.

**Three Dimensional (3D) Reconstruction**

The 2D image stack collected during the serial sectioning procedure was reconstructed in to a 3D object using the software package termed MIPAR™ (Materials Image Processing and Automated Reconstruction) based on a MATLAB platform, developed at The Ohio State University, and the commercially available AvizoFire 6.3 software. The entire 3D reconstruction procedure consists of (a) alignment of the stack, (b) cropping of a region of interest, (c) noise filtering/ thresholding, (d) segmentation, and (e) visualization. The first 4 steps were performed using MIPAR™ and AvizoFire 6.3 was used for 3D visualization. In MIPAR™, an FFT filter was applied to minimize curtaining effects (artifacts) obtained from FIB serial sectioning. Thresholding of TiC and graphite phases was done by applying a Gaussian blur subtraction along with the global range thresholding filter prior to segmentation of individual features. Finally, this entire post-processed 2D image stack was exported to AvizoFire 6.3 format for 3D visualization of TiC and graphite reinforcements.

## Results and Discussion

Figure 2(a) shows the x-ray diffraction (XRD) pattern for the as-deposited Ni-Ti-C composites of nominal compositions Ni-10Ti-10C and Ni-3Ti-20C. In case of Ni-10Ti-10C (lower pattern in Figure 2(a)), the primary diffraction peaks can be consistently indexed based on the face centered cubic (FCC) Ni phase and the δ-TiC phase exhibiting the rocksalt (NaCl type) structure. In case of Ni-3Ti-20C (top pattern in Figure 2(a)), a peak at $2\theta \sim 26°$, corresponding to the {0002} planes of graphitic carbon is clearly visible in addition to the FCC Ni and δ-TiC peaks. The absence of peaks corresponding to any Ni-Ti intermetallic phases is also evident in these diffraction patterns indicating that there is no reaction between nickel and titanium, rather titanium and carbon react within a molten nickel pool to form TiC precipitates.

Backscattered SEM images from the Ni-10Ti-10C composite are shown in Figures. 2 (b & c), clearly exhibiting the presence of TiC precipitates with two different morphologies and size- scales. The coarser and faceted carbides are likely to be the primary TiC precipitates while the finer scale needle-like carbides are likely to be eutectic TiC precipitates, homogeneously distributed within the nickel matrix. Backscattered SEM images of the Ni-3Ti-20C composite are shown in Figures 2 (d & e). These images clearly show the presence of TiC precipitates (grey) along with a substantial volume fraction of a phase exhibiting a black contrast, presumably corresponding to the graphitic phase revealed in the x-ray diffraction pattern. Ni-10Ti-10C exhibits approximately 17% (area fraction) of TiC whereas Ni-3Ti-20C exhibits approximately 8% graphite and 4% TiC. These 2D SEM images shown in Figures 2(b-e) give an idea of the overall microstructure of these composites and typically would be representative of microstructural analysis done in the past on such materials. However, the focus of the present study is a more detailed 3D analysis of the microstructure via serial-sectioning in dual-beam FIB and subsequent reconstruction of the 3D volume, as discussed below.

The reconstructed 3D volume corresponding to the Ni-10Ti-10C composite is shown in Figure 3(a). A cropped version of the entire 3D volume is shown in Figure 3(b) for better visualization purposes. In the cropped view, the primary TiC precipitates have been colored in green, while the eutectic TiC are represented in purple color. While the appearance of this 3D microstructure is rather complex it clearly highlights the following salient aspects of this microstructure:

a. The primary TiC precipitates exhibit a cuboidal morphology.
b. The eutectic TiC precipitates appear to exhibit a plate-like morphology in most cases unlike the needle-like morphology revealed by 2D SEM observations. Some needle-like eutectic TiC precipitates are also visible in the 3D reconstruction.
c. Nearly all the eutectic TiC precipitates are connected to the primary TiC as well as other neighboring eutectic TiC precipitates.
d. The eutectic TiC precipitates form a 3D network with the larger primary TiC precipitates located at the nodes of this network.

The connectivity between the carbide precipitates is impossible to discern based on the 2D SEM images in Figure 2, without the aid of 3D reconstruction. The eutectic TiC precipitates that are connected to the primary TiC, possibly nucleated from the primary precipitate during solidification.

Figure 4(a) shows the 3D reconstruction of the Ni-3Ti-20C composite. The major challenge in 3D reconstruction of these Ni-3Ti-20C composites compared to the Ni-10Ti-10C composites is the presence of 2 different precipitates phases i.e. TiC and graphite. During post processing of 2D SEM images, both the phases (TiC and graphite) were thresholded differently and then exported to reconstruct the 3D volume. The TiC precipitates are yellow in color while the graphite bundles are pink colored. The carbide precipitates (yellow) are clearly connected to each other and form a complex network in 3D. This connectivity between the carbide precipitates is nearly impossible to visualize based on the 2D SEM images shown in Fig. 2. The distinction between the cuboidal primary TiC precipitates and the plate or needle-shaped eutectic TiC precipitates is more difficult in case of this composite since the primary precipitates are smaller in size as clearly shown in the higher magnification view of a cropped section of the 3D reconstruction in Figure 4(b). Furthermore, most of the cuboidal primary TiC precipitates are engulfed by graphite bundles (pink) as shown in both Figs. 4(a and b). Referring to the 2D SEM images shown in Figures 2(d and e), there is a gap between the TiC precipitates and the graphite bundles, and it is very difficult to comment on their interconnectivity and association based on these images. The 3D reconstructions shown in Figure 4 conclusively reveal the true nature of the connection between the carbide precipitates and the graphitic bundles in this microstructure. Referring to Figures 4(a and b), the eutectic TiC precipitates in case of the Ni-3Ti-20C composite exhibit both plate-like and needle-like morphologies

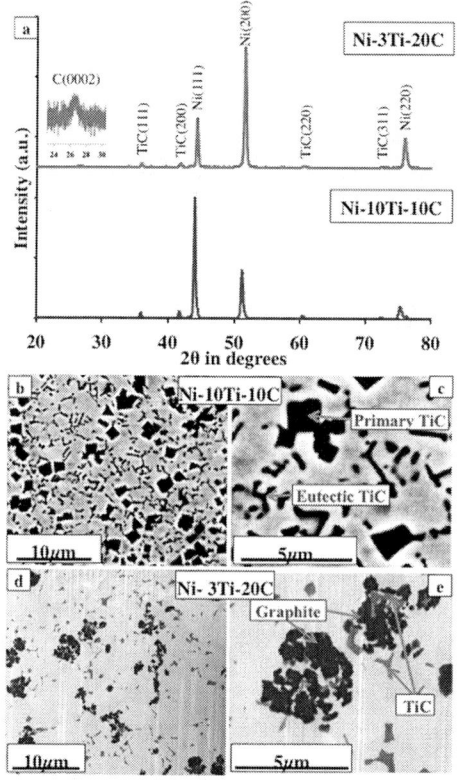

**Figure 2.** (a) XRD pattern obtained from the LENS deposited Ni-Ti-C composites. (b & c) Backscatter SEM images of LENS deposited Ni-10Ti-10C composites (d & e) Backscatter SEM images of LENS deposited Ni-3Ti-20C composites.

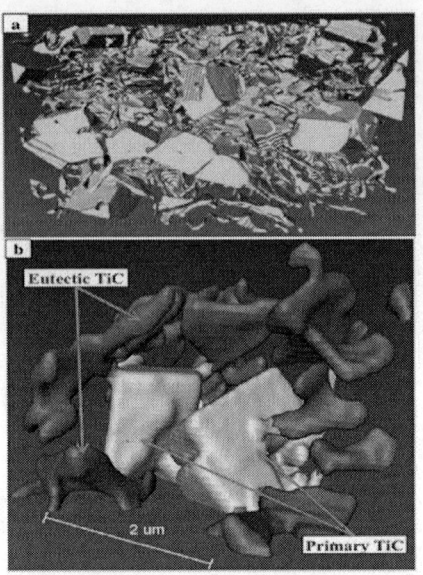

**Figure 3.** (a) 3D reconstruction of Ni-10Ti-10C composites obtained from the 2D SEM image stack
(b) Cropped version of 3D reconstructed volume of Ni-10Ti-10C showing primary and eutectic TiCs.

The 3D reconstructions aid in developing a better understanding of the sequence of phase evolution during solidification for both types of composites (Ni-10Ti-10C and Ni-3Ti-20C) as well as the shape, size, and connectivity between the different phases involved. Within the liquid melt pool consisting of Ni+Ti+C, the primary TiC forms appears to be the first solid phase to form during solidification for both types of composites, and at lower temperatures, the eutectic TiC as well as graphitic phases are formed (only for Ni-3Ti-20C). For both types of composites, the cuboidal primary TiC precipitates appear to act as the heterogeneous nucleation sites for the eutectic product. Thus, in case of the Ni-10Ti-10C composite, the primary TiC cuboids act as the nodes from which the eutectic TiC precipitates (plate-like and needle-like) nucleate resulting in the formation of the complex 3D interconnected network of carbide precipitates. In case of the Ni-3Ti-20C composite, smaller cuboidal primary TiC precipitates form in the liquid and these act as heterogeneous nucleation sites for both the eutectic TiC precipitates as well as graphite bundle that form presumably at lower temperatures and engulf the primary TiC precipitates. While these insights into the solidification sequence could not have been revealed without the 3D reconstructions, it should be noted that more detailed investigations are warranted in order to develop a robust understanding of the microstructural evolution process. The volume fractions of the different phases have also been calculated using the 3D reconstructions. The results indicate that Ni-10Ti-10C exhibits approximately 17% of TiC whereas Ni-3Ti-20C exhibits approximately 7%.

graphite and 5% TiC which is in very good agreement with the values computed from the 2D area fractions.

Comparing the microhardness values for the two different composites, the Ni-10Ti-10C composite exhibited a substantially higher hardness of 370 VHN as compared to 165 VHN for the LENS$^{TM}$ deposited pure Ni and 290 VHN & 240 VHN for LENS deposited Ni-7Ti-20C & Ni-3Ti-20C respectively. These microhardness values have been listed in Table I and clearly show a trend of decreasing hardness as a function of increasing C/Ti ratio in the composite.

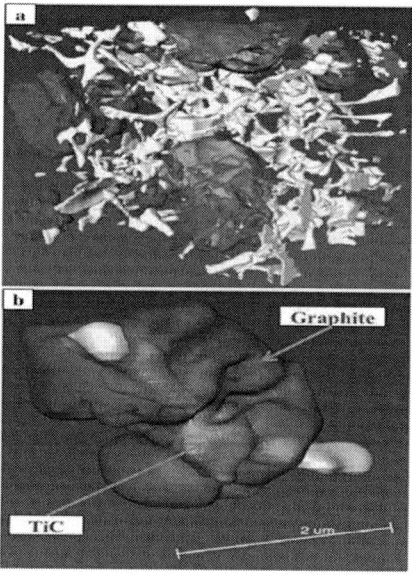

**Figure 4.**  (a)  3D reconstruction of Ni-3Ti-20C composites obtained from the 2D SEM image stack
(b)  Cropped version of 3D reconstructed volume of Ni-3Ti-20C showing primary and eutectic TiC engulfed by graphite bundle.

The high volume fraction of titanium carbides (both primary and eutectic), in case of the Ni- 10Ti-10C composite resulted in the higher microhardness value. Preliminary results of POD tribometry studies carried out on the LENS deposited Ni-10Ti-10C, Ni-3Ti-20C, Ni-7Ti-20C composites and pure Ni samples are shown in Figure 5

| Composite Sample | Hardness(HV) |
|---|---|
| Pure Ni | 165 ± 6 |
| Ni-3Ti-20C | 240 ± 6 |
| Ni-7Ti-20C | 290 ± 7 |
| Ni-10Ti-10C | 370 ± 10 |

Table I. The microhardness values for the different Ni-Ti-C composites investigated in the present study in comparison with pure Ni

From Figure 5, it is clear that the graphite and TiC phases in the composite were beneficial in reducing the friction coefficient with respect to the pure Ni sample. While the presence of TiC reduces the coefficient of friction, as observed in case of the Ni-10Ti-10C composite, the presence of the lubricious graphitic phase can play a more dominant role in reducing the friction for these composites. This is evident from the friction curves for the Ni-7Ti-20C and Ni-3Ti-20C composites. The friction coefficient in case of Ni-7Ti-20C is marginally lower as compared to Ni-10Ti10C due to the presence of the graphitic phase in the former. However, the most promising composite appears to be the Ni-3Ti-20C composite which exhibits a drastic reduction in friction coefficient (~0.2) as compared to any of the other composites, mainly due to the presence of a substantial fraction of the graphitic phase as well as TiC precipitates. Thus, these Ni-Ti-C composites, especially the Ni-3Ti-20C composite, appear to be promising materials for surface engineering applications requiring high hardness with improved solid lubrication

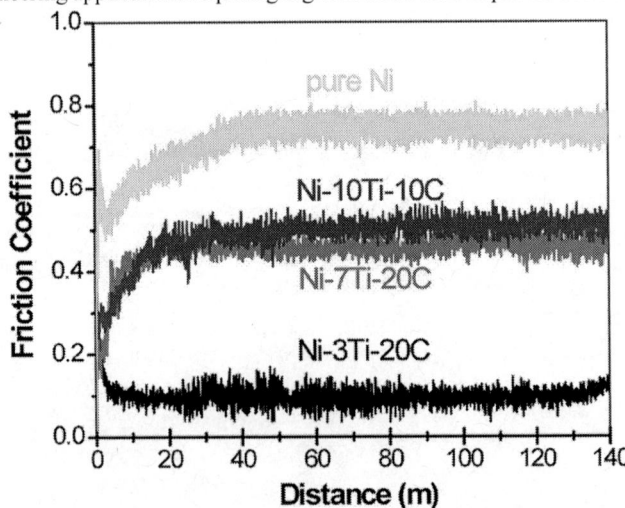

Figure 5. Steady state friction coefficient as a function of sliding distance up to 140 m for LENS deposited pure Nickel, Ni-10Ti-10C, Ni-7Ti-20C, and Ni-3Ti-20C composites.

## Conclusions

Novel in situ Ni-Ti-C composites have been deposited using the laser engineered net shaping (LENS$^{TM}$) process. While the Ni-10Ti-10C composite exhibits a large volume fraction of primary cuboidal TiC precipitates as well as eutectic carbide precipitates, both the Ni-7Ti-20C and Ni-3Ti-20C composites exhibit an additional graphitic phase. 3D microstructural characterization of these composites reveals the following salient features of the these composites:

1. The Ni-10Ti-10C composite consists of a complex interconnected network of carbide precipitates with primary cuboidal TiC precipitates at the nodes of the 3D network, connecting plate-like (needle-like in some cases) eutectic TiC precipitates. The primary carbides appear to act as the heterogeneous nucleation sites for the eutectic carbides.

2. The Ni-3Ti-20C composite also consists of an interconnected network of primary carbide precipitates at nodes with eutectic carbides connecting them, but the smaller scale cuboidal primary carbide precipitates in this case are engulfed by graphitic bundles, a second primary solidification product.

Such 3D characterization leads to a better understanding of the sequence of phase evolution during solidification in these novel composites. tribological and mechanical property measurements reveal that the steady-state friction coefficients for these Ni-Ti-C composites are significantly lower than that of pure Ni, while their microhardness is substantially higher, making them promising candidates for surface engineering applications that require both solid lubrication and high mechanical hardness. Out of all the Ni-Ti-C composites investigated the Ni- 3Ti-20C composite appears to exhibit the best balance of properties with the lowest coefficient of friction as compared to pure Ni, Ni-10Ti-10C, and Ni-7Ti-20C, and a microhardness that is greater than pure nickel, though being somewhat lower than Ni-10Ti-10C or Ni-7Ti-20C composites.

## Acknowledgements

This work has been supported by the ISES contract awarded to the University of North Texas by the U.S. Air Force Research Laboratory, AFRL contract number FA8650-08-C-5226. The authors also gratefully acknowledge the Center for Advanced Research and Technology (CART) at the University of North Texas.

## References

1. S. R. Bakshi, D. Lahiri, A. Agarwal, *International Materials Reviews*, 55, 41-64 (2010).

2. N. Saheb, Z. Iqubal, A. Khalil, A. Hakeen, N. Aqeel, T. Laoui, A. Qutum, R. Kirchner, *Journal of Nanomaterials*, 2012, 1-13 (2012).

3. S. C. Tjong, Z. Y. Ma, *Materials Science Engineering*. R, 29, 49-113 (2000).

4. D. B. Miracle, *Composites Science and Technology*, 65, 2526-2540 (2005).

5. S. Rawal, *Journal of Metals*, 53, 14-17 (2001).

6. S. Cho, K. Kikuchi, T. Miyazaki, K. Takagi, A. Kawasaki, T. Tsukada, *Scipta Materialia*, 63, 375-378 (2010).

7. K. Zhuang, L. Wang, Nanosci. *Nanotechnology Letters*, 4, 454-456 (2012).

8. S. Gopagoni, J. Y. Hwang, A. R. P. Singh, B. A. Mensah, N. bunce, J. Tiley, T.W. Scharf, R. Banerjee, J. Alloys and Compounds, 509, 1255-1260 (2011).

9. X. Wang, A. Jha, R. Brydson, *Materials Science Engineering* A, 364, 339-345 2004).

10. Z. Liu, J. Tian, B. Li, L. Zhao, *Materials Science Engineering* A, 527, 3898-3903 (2010).

11. D. Strzeciwilk, Z. Wokulski, P. Tkacz, J. Alloy Compd., 350, 256-263 (2003). G. Xiao, Q. Fan, M. Gu, Z. Wang, Z. Jin, *Materials Science Engineering* A, 382, 132-140 (2004).

13. Y. Li, P. Bai, Y. Wang, J. Hu, Z. Guo, *Materials Design*, 30, 1409- 1412 (2009).

14. D. Strzeciwilk, Z. Wokulski, P. Tkacz, *Crystal Research Technology*, 38, 283-287 (2003).

15. M. D. Uchic, M. Groeber, D. Dimiduk, J. P. Simmons, *Scripta Materialia*, 55, 23-28 (2006).

16. M. A. Grober, B. K. Haley, M. D. Uchic, D. M. Dimiduk, *Materials Characterization*, 57, 259-273 (2006).

17. R. T. DeHoff, *Journal of Microscopy*, 131, 259-263 (1983).

18. D. N. Dunn, R. Hull, *Applied Physics Letters*, 75, 3414-3416 (1999).

19. B. J. Inkson, M. Mulvihill, G. Mobus, Scripta Materialia, 45, 753-758 (2001).

20. P. Burdet, J. Vannod, A. Hessler-Wyser, M. Rappaz, M. Cantoni, Acta Mater., 61, 3090- 3098 (2013)

21. G. Gaiselmann, M. Neumann, L. Holzer, T. Hocker, M. R. Prestat, V. Schmidt, Compsite Materials Science, 67, 48-62 (2013).

22. K. Scott, Journal of Microscopy, 242, 86-93 (2011).

# USE of SQUEEZE INFILTRATION PROCESSING for FABRICATING MICRO SILICA REINFORCED ALUMINUM ALLOY-BASED METAL MATRIX COMPOSITE

K. M. Sree Manu, V.G. Resmi, T.P.D. Rajan, Prince Joseph and B.C. Pai

*Materials Science and Technology Division,*
CSIR-*National Institute for Interdisciplinary Science and Techn*ology
Thiruvananthapuram 695019, **India**

T.S. Srivatsan

Department of Mechanical Engineering
**The University of Akron**
Ohio 44325-3903, **USA**

## Abstract

A number of processing techniques have been developed and tried for the primary purpose of synthesizing metal-matrix composites (MMCs) using a variety of alloy and reinforcement combinations. In particular, the technique of infiltration processing has been preferred and used for making composites based on metal matrices and reinforced with a high volume fraction of the reinforcing phase. In this paper, the results of a study on using fine particle size microspheres as the reinforcement for an aluminium alloy metal matrix is presented and discussed. Polyethylene glycol was used as the pore former along with aluminium powder as the binding agent. The resultant mixture was sintered at a high temperature. The preforms were compacted and subsequently sintered at a temperature above the melting temperature of aluminium. Decomposition of the fugitive pore former was done using the burnout technique. Key aspects pertinent to characterizing the fabricated preforms and composite using the technique of light optical microscopy, thermo gravimetric analysis (TGA) and hardness measurement is briefly presented.

**Keywords:** micro-silica preform, squeeze infiltration, metal-matrix composites, aluminium alloy.

## 1. Introduction

Both the existing and rapidly emerging aerospace systems have mission requirements that will demand the advanced materials, such as reinforced metal matrices, to have the potential to not only meet but even exceed the demands set by the era of prevailing aerospace and emerging aerospace vehicles. Most noticeably during the last two decades, since the early 1990s, reinforced metal matrices have emerged as a promising, potential, and a viable materials solution for emerging structures used in both high performance-critical and non-performance-critical applications. Reinforced metal matrices can be considered to be versatile engineering materials since they offer the potential for noticeable improvements in efficiency, stiffness and strength coupled with not only good wear and abrasion resistance but also acceptable corrosion resistance over the newer and emerging generation of: (a) the monolithic alloys of aluminium and titanium, and (b) the organic-matrix composites [1-7]. However, the primary disadvantage with these materials is that they tend to frequently suffer from an observable combination of low tensile ductility (elongation-to-failure), inadequate fracture toughness, and inferior fatigue and fracture resistance when compared one-on-one with the unreinforced matrix alloy [8-16]. Further, reinforced metal matrices have the innate ability of being processed and finished using conventional metal processing techniques and facilities. Most importantly and/or noticeably during the last two decades the technology of discontinuously-reinforced aluminium alloy-based composites has matured to the point where both critical and non-critical components are now being produced on a commercial scale for use in aircraft structures, gas turbine engines, automobiles, electronics, spacecraft, and even recreational goods [17-19].

In the specific domain of processing sustained research and development efforts have culminated in the emergence and use of the infiltration process as a promising and viable alternative approach for the production of metal ceramic composites containing a relatively high volume fraction of the reinforcing phase in the metal matrix [20]. The resultant composite experiences noticeably less shrinkage coupled with reduced gas porosity. This is because solidification is promoted to occur under the influence of applied pressure resulting in an end product that is likely to have improved properties, spanning both mechanical and physical, to offer. This technique has gradually grown, both in stature and significance, to gain attention for purpose of selection and eventual use in related industry circles for the production of polymer-based, ceramic-based and even metal-based composites. With specific reference to infiltration the key difference between any two methods is based entirely on the technique that is used to cause the molten metal to enter the preform.

Synthesis of self-sustaining porous ceramic preform having sufficient strength and porosity is a key step in the infiltration processing of composites. The ceramic foams and preforms also possess a wide variety of other applications due to a combination of attractive properties they have to offer, such as, (i) thermal and corrosion resistance, (ii) low density, (iii) low thermal conductivity, (iv) controlled permeability, (v) high surface area, and (vi) high structural uniformity [21, 22]. A few of the important fabrication methods that have been tried and used for porous ceramic foams are: (i) polymer replica techniques, (ii) gel casting, (iii) direct foaming of suspensions, and (iv) use of pore forming agents (PFA) [23, 24]. The preform route does offer a wide variety of types, morphologies and metal volume content. Thus, tailored microstructures having an interpenetrated network can be easily realized. Properties of the porous ceramic and the metal melt coupled with their mutually interactive interactions are the key factors that exercise control over properties of the engineered composite.

The present investigation was based on synthesizing micro silica preform using polyethylene glycol as the pore former, aluminium as the binder and direct squeeze infiltration of the molten aluminium alloy to the porous preform to fabricate the Al alloy-micro silica metal ceramic composite. Direct squeeze infiltration of the chosen aluminium alloy onto the preform was successfully carried out using controlled process parameters spanning
(i) Initial preform temperature
(ii) Liquid metal superheat
(iii) Squeeze pressure and its rate of application, and
(iv) Temperature of the die.

## 2. Materials and Experimental Methods

### 2.1 Material and Fabrication

Aluminium alloy 319 (referred to henceforth as AA319) was chosen to be the matrix phase for purpose of infiltration. The nominal chemical composition (in weight percent) of tis alloy is 6.35 Si, 3.87 Cu, 0.20 Fe, other trace elements, and balance Al. The macro porous performs were prepared by stacking a powder mixture of micro-sized silica, polyethylene glycol and aluminium powder. The role of aluminium powder is that of the binding agent. The uniformly mixed powder mixture was then transferred to a steel mold having a diameter of 25 mm and subsequently precision compacted at a pressure of 45 MPa, using a CARVER press, and at room temperature (25°C). The compacted preforms were subsequently fired at a controlled heating rate so as to minimize disruption or cracking during firing. The ceramic preforms were heated in a furnace at a temperature above the melting point of aluminium with the primary objective of promoting the aluminium to act as a binder. During the process of heating decomposition of the polyethylene glycol takes place resulting in the creation of fine pores in the preform.

Maintaining the process parameters helps in successive infiltration of the metal matrix for the purpose of engineering the composite. The infiltration die, which is made up of tool steel, was heated to $300^0C$. This was followed by pre-heating the micro silica preform to $600^0C$ and is then placed in the die prior to pouring of the liquid molten metal. The super-heated aluminium alloy was poured into the die in which the preform was placed. Subsequently, high pressure was applied on the molten metal to facilitate its infiltration into the preform. This was made possible using a hydraulic press. The pressure was maintained for about 5-6 minutes until such time the solidification was complete.

## 2.2 Microstructural Characterization

Samples of the engineered composite were prepared very much in conformance with standard procedures used for the metallographic preparation of metal samples. This involved coarse polish using progressively finer grades of silicon carbide (SiC) impregnated emery paper [i.e., 320-, 400- and 600-grit] followed by fine polishing using an alumina-based polishing compound suspended in distilled water as the lubricant. The polished samples were examined in an optical microscope, at low magnifications, and photographed using standard bright field illumination technique.

## 2.3 Thermogravimetric analysis (TGA)

Also referred to as Thermalgravimetrix Analysis (TGA) is a method that is often used to assess the changes in both physical and chemical properties of materials as a function of increasing temperature, often at a constant loading rate or as a function of time. The TGA analysis or measurement does provide useful information about physical phenomena, such as vaporization, sublimation, absorption adsorption and desorption. In this study, TGA was used to establish the mass loss or gain experienced by the polymeric material used either due to decomposition, oxidation or loss of volatile elements, such as moisture. The results obtained using the TGA instrument are plotted with temperature (in °C) on the X-axis and mass loss (in mg) on the Y-axis. The data is often adjusted using curve smoothing and the points of inflection are determined and recorded.

## 2.4 Hardness Testing

A basic mechanical property of a material is its hardness. The hardness test is an important and often most widely used test for the purpose of evaluating the mechanical properties of monolithic metals, their alloy counterparts, and even composite materials based on metal matrices. A simple yet appropriate definition that has been recorded for hardness is the resistance offered by the material to indentation, i.e., permanent deformation and cracking [25]. A direct measurement of hardness is both a simple and useful technique for characterizing the base-line mechanical properties while concurrently investigating, establishing and rationalizing the role and contribution of intrinsic microstructural constituents. Overall, the hardness test can be safely considered to be both simple and easy enough to enable it to be safely categorized as being non-destructive [25, 26]. In this study, the Brinell hardness ($H_{BH}$) measurements were made on a macrohardness tester using an indentation load of 60 kgf, a dwell time of 15 seconds, with the aid of a 2.5 mm diameter steel ball. The indenter rests for a specified length of time on the polished surface of the test specimen. The machine makes an indent, or impression, on the polished surface of the sample whose diagonal size was measured using a low magnification optical microscope. The area of the impression is directly proportional to the load used and a load independent hardness number can be found. The Brinell hardness number ($H_{BHN}$) is the ratio of applied load to the surface area of the indent and was provided by the test machine. The result is reported as the average value in units of $kg/mm^2$.

## 3. Results and Discussion

In Figure 1a reveals the sintered micro silica porous preform synthesized using the by burnout technique with polyethylene glycol as the pore former. A quality preform forms the basis for a high-quality metal-matrix composite. Strength of the preform is largely dependent on the kind of binder used during fabrication. In this research study, aluminium was used as the binder, which not only aids in stabilizing the shape but also provides sufficient strength to the preform. The fabricated preform should satisfy both the mechanical and thermal properties so as to prevent deformation of the preform during the process of infiltration. In Figure 1 (b) is shown the macrograph of the preform infiltrated specimen. It is very clear that the fabricated preform has good "open" porosity, which results in successive infiltration by the molten metal, i.e., liquid aluminium alloy. This macrograph reveals the infiltration process to be a suitable method for the fabrication of selectively reinforced composites.

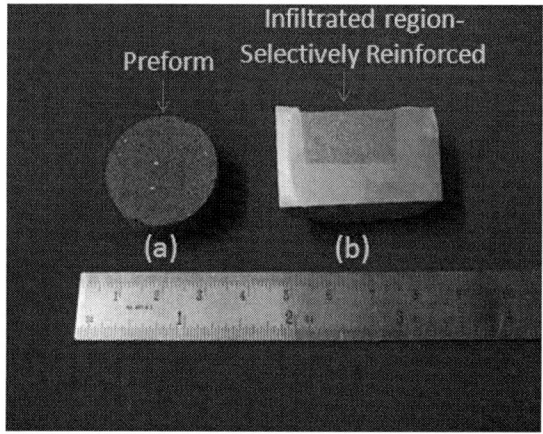

Figure 1. (a) Macrograph of the sintered porous preform, and
(b) The infiltrated composite

The microstructure of the aluminium alloy composite, i.e., AA319 +50 vol. % micro silica, made by the method of squeeze infiltration is shown in Figure 2. Macroscopic or low magnification observation revealed a near uniform distribution of the reinforcing micro-silica particles in the aluminium alloy metal matrix. The preform was completely infiltrated by the molten aluminium alloy even through minute pores. In discontinuous particulate-reinforced aluminium alloy-based composites the most important factor is achieving a near uniform dispersion of the reinforcing phase in the metal matrix, which decides the final mechanical properties of the engineered composite material. Light optical microscopy observations revealed the microstructure to have a high volume fraction of the reinforcing particulates or particles in the aluminium alloy metal matrix coupled with small quantities of agglomeration of the pore forming agent resulting in the creation of large pores that facilitate penetration by the bulk metal. However, using the squeeze infiltration process it appears that a high volume fraction of the reinforcing phase (i.e., particulates) can be near uniformly distributed through the metal matrix. In Figure 3 is shown the microstructure at the region of the interface between the matrix, i.e., aluminium alloy 319 and the micro silica reinforcing particles. At the interface region no reaction zone was observed due essentially to a good penetration of the liquid metal, which depends on the infiltration parameters used. The parameters include the following: (i) preform temperature, (ii) die temperature, (iii) squeeze pressure, and (iv) temperature of the liquid aluminium alloy.

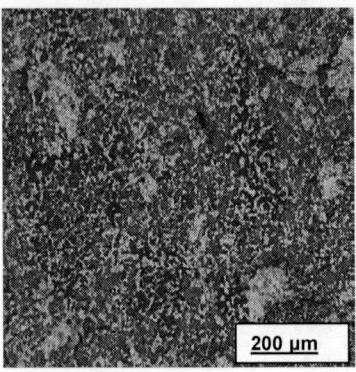

**Figure 2.**   Optical micrograph showing microstructure of aluminium alloy 319 reinforced with 50 vol. % micro silica

**Figure 3.**   Light optical micrograph showing microstructure at the interface region between the aluminium alloy matrix and the reinforcement.

The thermo-gravimetric analysis (TGA) curve of the preform having equal vol. % of micro silica and polyethylene glycol is shown in the Figure 4. The TGA spectrum of the preform was done as a function of temperature; at a heating rate of $10°$ C/min in environment of ambient air ($27°$ C). Thermal decomposition of the polyethylene glycol started at $200^0$ C and ended at around $260^0$ C.

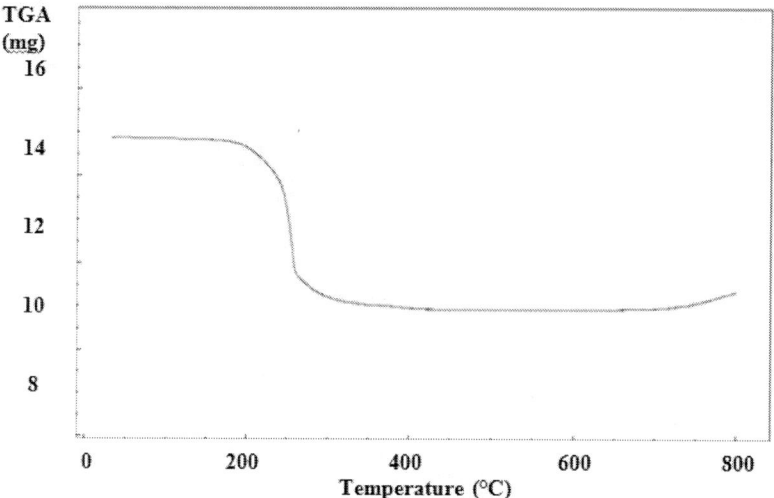

**Figure 3.** TGA curve of the porous preform having equal vol.% of micro silica and polyethylene glycol.

The Brinell hardness values for aluminium alloy 319 and sample of the engineered metal matrix composite fabricated using the technique of squeeze infiltration processing are listed in Table.1. Presence of the reinforcing phase did increase the hardness of the matrix alloy. The results obtained reveal the composite microstructure to have noticeably higher hardness than the unreinforced metal matrix. Hardness of the metal matrix, i.e. AA319 is 95 BHN, while macro-hardness of the engineered composite is 155. An overall sixty five percent improvement in macro-hardness

**Table 1.** Brinell hardness values for sample processed using the squeeze infiltration process.

| Sample Type | Hardness (BHN) |
|---|---|
| Alloy | 95 |
| Composites | 155 |

## 4. Conclusions

The following are the essential findings of this research study:

1. Liquid metal infiltration of ceramic preform is a promising route to produce metal matrix composites having a high volume fraction and near uniform distribution of the reinforcing phase in the metal matrix.

2. The technique of burn out of polyethylene glycol (PEG) was used to fabricate macro porous preform for the purpose of liquid metal infiltration.

3. The Thermo-gravimetric analysis (TGA) curve reveals decomposition of the PEG to occur in the range $200^0$ C - $260^0$ C.

4. Direct squeeze infiltration of aluminium alloy 319 onto the micro silica preform was successfully carried out by following the process parameters till such time the molten aluminium alloy completely infiltrated through the preform.

5. The infiltrated AA319-micro silica composite revealed a near uniform distribution of the reinforcing phase in the metal matrix resulting essentially in a microstructure that was free of pores.

6. Macro hardness of the AA 319– micro silica composite was 155 BHN when compared to 95 BHN for the base alloy in as-cast condition.

**Acknowledgements**

The authors are grateful to the *Director* and Members of *Materials Science and Technology Division*, CSIR-NIIST, Trivandrum, India for their support and help.

# References

1. R. H. Van Stone, T.B. Cox, J.R. Low Jr., and J.A. Psioda: *International Metals Reviews*, Vol. 30, 157, 1975.

2. S.V. Nair, J. K. Tien and R. C. Bates: *International Metals Reviews*, Vol. 30, No. 6, p. 285, 1985.

1. M. Taya and R. J. Arsenault: *Metal Matrix Composites*, Pergamon Press, Elmsford, New York, 1989.

2. W. H. Hunt, Jr., C.R. Cook and R. R. Sawtell SAE Technical Paper Series 91-0834, Society of Automotive Engineers, Warrendale, PA, USA, 1991.

3. T.S. Srivatsan, T. S. Sudarshan and E. J. Lavernia, in *Progress in Materials Science*, Vol. 39, 1995, p. 317.

4. B. Marayuma: The AMPTIAC Newsletter, Vol. 2, No. 3, 1998.

7. A.S. Argon, J. Im and R. Safoglu: *Metallurgical Transactions*, Vol. 6A, 825, 1975.

8. P.K. Liaw and W. A. Logsdon: *Engineering Fracture Mechanics*, Vol. 24, 1986, pp. 737-745

9. J.K. Shang and R. O. Ritchie: *Metallurgical Transactions*, Vol. 20A, 1989, opp. 897-908.

10. J.K. Shang and R. O. Ritchie: *Acta Metallurgica*, Vol. 37, 19089, pp. 2267-2278

11. D.L. Davidson: *Journal of Materials Science*, Vol. 24, 1989, pp. 681-687.

12. M. Manoharan and J. J. Lewandowski: *Acta Metallurgica*, Vol. 38, No. 3, 1990, pp. 489-496.

13. D.L. Davidson: *Engineering Fracture Mechanics*, Vol. 33, 1989, pp. 965-977.

14. J.J. Lewandowski, C. Liu and W.E.H. Hunt, Jr.: *Materials Science and Engineering*, Vol. 107A, 1989, pp. 49-55.

15. M. Manoharan and J. J. Lewandowski: *Scripta Metallurgica*, Vol. 23, 1989, pp. 301-305.

16. M. Vogelsang,. R. J. Arsenault and R. M. Fisher: *Metallurgical Transactions*, Viol. 17A, 1986, pp. 379-388.

17. R.J. Arsenault in *Metal Matrix Composites: Mechanisms and Properties* (editors: R.K. Everett and R.J. Arsenault), Academic Press, San Diego, CA, 1991, pp. 79-87.

18. K.K. Chawla and M. Metzger: *Journal of Materials Science*, Vol. 17, 1972, pp. 34-43

19. M. Taya and T. Mori: *Acta Metallurgica*, Vol. 35, 1987, pp. 155-163.

20. D. Huda, M.A. El Baradie and M.S. Hashmi, *Journal of Materials Processing Technology*, 37, (1993) pp.513-528

21. M. Scheffler, P. Colombo, *Weinheim, Wiley-VCH*, (2005) pp. 645.

22. Paolo Colombo, *Philosophical Transactions Of The Royal Society A*, 364 (2006) 109-124.

23. P. Sepulveda, *American Ceramic Society Bulletin*, 10 (1997) 61–65.

24. L. Montanaro, Y. Jorand, G. Fantozzi, A. Negro, J. Eur. Ceram. Soc. 18 (1998) 1339-1350.

25. D. R. Askeland, and P. P. Phulke, The Science and Engineering of Materials, Fifth Edition, Thompson Publishers, USA, 2005, 520-530, 2008.

26. G.E. Dieter, G.E, Mechanical Metallurgy, Third Edition, McGraw Hill Publishers, 400-410, 2006.

# FABRICATION AND CHARACTERIZATION OF A HYBRID FUNCTIONALLY GRADED METAL-MATRIX COMPOSITE USING THE TECHNIQUE OF SQUEEZE INFILTRATION

K. M. Sree Manu[**], V.G. Resmi, Prince Joseph, T.P.D. Rajan, and B.C. Pai

*Materials Science and Technology Division,*
CSIR-National Institute for Interdisciplinary Science and Technology
Thiruvananthapuram 695019, **India**

T.S. Srivatsan

Department of Mechanical Engineering
**The University of Akron**
Ohio 44325-3903, USA

## Abstract

In this research paper, the results of a recent study that focussed on preparing hybrid and graded cylindrical preforms by stacking alumina-silicate fibre, SiC particles and the pore forming agent in different layers along with aluminium powder as the binding agent and the resultant mixture sintered at a high temperature is presented and discussed. The stacked preforms were compacted and sintered above the melting temperature of aluminium. Decomposition of the fugitive pore former was done using the leaching technique. Direct squeeze infiltration of aluminium alloy 319 on a graded hybrid preform was successfully carried out using controlled process parameters spanning initial preform temperature, liquid metal superheat, squeeze pressure and its rate of application, and die temperature. The liquid metal did fully infiltrate through the ceramic preform to form a functionally graded composite. The resultant composites were characterized using the techniques of scanning electron microscopy, optical microscopy, x-ray diffraction and measurement of density. The extrinsic influence of infiltration pressure, preform temperature and mold temperature on microstructural development and hardness of the engineered composites is briefly highlighted.

**Key words:** Squeeze Infiltration, Ceramic Preform, hybrid metal matrix composites, Aluminium alloy, Alumino-silicate

---

**corresponding author:  K.M. Sree Manu
Email:                  manu.kaimanikal@gmail.com

## Introduction

Pressure infiltration of liquid metal into a ceramic preform can be safely categorized to be one of the commonly used fabrication methods for the production of metal matrix composites having a high volume fraction of reinforcement [1-3]. Squeeze infiltration is one of the techniques that have been successfully used to obtain a high volume fraction of reinforcement in a functionally graded material. Graded porous ceramics serve as the starting preform for the fabrication of functionally-graded materials using the technique of liquid metal infiltration. The significant advantages offered by the squeeze infiltration process over the conventional methods used for fabrication, such as casting, is by overcoming the common problems associated with the following: (i) poor wettability between the matrix and the reinforcement, (ii) extensive interfacial reactions, (iii) non-uniform distribution of the reinforcement in the metal matrix, (iv) shrinkage, and (v) gas porosity [4]. Metal-ceramic composites having a high volume fraction of the hard, brittle and elastically deforming ceramic reinforcement in the soft, ductile and plastically deforming metal matrix have in recent years grown both in stature and strength to enable its selection and use in the following fields: (i) automotive and electronic packaging in the role of heat spreader, (ii) brake pads, (iii) diesel piston, and (iv) heat shields, to name a few. In particular, these composites are noted for their high strength, high stiffness, superior wear resistance, coupled with enhanced high temperature properties.

A self-sustaining porous ceramic preform serves as the starting material for the fabrication of a metal-matrix composite (MMC) by the technique of liquid metal infiltration. Over the years, a wide range of processing routes has been proposed for the production of porous ceramics. A few of these include the following:
(a)   Polymer replica technique,
(b)   Direct foaming of liquid slurry,
(c)   The burn out technique, and
(d)   Leaching method.
The end result is a variety of morphologies [5]. For a hybrid ceramic preform some of the common problems that tend to occur during manufacturing are: (i) the non-uniform arrangement of the components within the preform structure, (b) agglomeration, and (c) closed porosity. Squeeze infiltration can be safely categorized as being a liquid state fabrication process. In this process, the molten metal gets penetrated by a preform of fibres, whiskers or a porous bed of loose particulates upon the application of pressure [6-9]. During squeeze infiltration the contact time between the melt and the reinforcing phase is often less. This helps in minimizing some of the brittle interfacial reactions that tend to occur with a concomitant degradation in the mechanical properties.

## Materials and Experimental Methods

The commercial aluminium alloy 319 (AA319) was used as the matrix material. The nominal chemical composition (in weight percent) is 6.35 Silicon, 3.87 Copper, 0.20 Iron, other trace elements, and remaining aluminium. The reinforcing phases used were (i) particulates of silicon carbide (SiCp) having an average size of 42μm, and (ii) alumina-silicate fibres. The fibres were acquired from M/s Murugappa Morgan Thermal Ceramics Ltd., (Chennai, India). The fibres synthesized by use of melt spin method were of assorted length and diameter. The average fibre diameter was in the range of 1.5-3 μm. The chemical composition of the fibre is summarized in **Table 1**.

| Constituents | Amount % |
|---|---|
| $Al_2O_3$ | 43-47 |
| $SiO_2$ | 53-57 |
| $Fe_2O_3$ | 0.02-0.08 |
| MgO | 0.01-0.04 |
| CaO | 0.05-0.4 |
| Others | Traces |

Table.1 Composition of the alumina-silicate fibres

Varying compositions of the macro-porous hybrid preforms were prepared by carefully and completely mixing the alumina-silicate short fibres, silicon carbide particles ($SiC_p$), sodium chloride (NaCl) and the aluminium alloy powder using a medium of acetone. The medium was subsequently evaporated at room temperature (25°C). Upon drying, the ceramic mixture was transferred to a steel mold having a diameter of 25-mm and subsequently compacted, using a CARVER press, at ambient temperature (25°C) using a pressure of 45-MPa. The compacted green body was then heated in a furnace at a temperature slightly above the melting point of aluminium alloy. The salt present in the pellet was removed by leaching in hot water. This was followed by drying at $120^0C$ so as to obtain a porous hybrid preform.

An infiltration die having a cavity that measured 35-mm in diameter was designed and fabricated using tool steel. The SiC preform was preheated to $650^0C$ and then placed in the die and heated at an elevated temperature of $180^0C$ prior to pouring the liquid molten metal. The molten metal containing magnesium was heated to $750^0C$ and then poured into the die containing the preform. High pressure was then applied to the molten metal in order to facilitate penetration into the porous hybrid preform. The pressure (MPa) was maintained for about 5–6 minutes till such time solidification was complete. A phase analysis of the samples was performed using X-ray diffraction [XRD, PW-1700, Philips Co.] with $CuK_α$ radiation. Surface morphology of the SiC preform was determined with the aid of a JEOL scanning electron microscope (SEM). Intrinsic microstructural features of the aluminium alloy metal matrix composite were observed with the aid of a Leica Optical Microscope. Macro-hardness measurements were made on both the metal-ceramic composites and the base material using a Brinell hardness tester.

### Results and Discussion

In **Figure1** is shown the macrograph of the three sinteredporous hybrid preforms and the infiltrated composite having: (a) 20 vol.% fibre, (b) 20 vol.% fiber-10 vol.% $SiC_p$, (c) 20 vol.% fibre – 20 vol.% $SiC_p$, and (d) AA 319 with 20 vol.% alumino-silicate fibre plus 20 vol.% $SiC_p$ hybrid preform composite synthesized using the NaCl leach out technique.

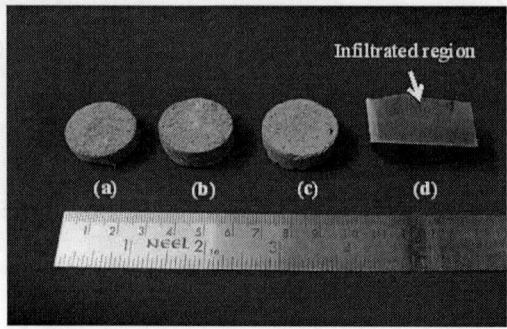

**Figure 1.** Macrograph of the three sintered porous hybrid preforms and the infiltrated composite

In **Figure 2(a)** is shown scanning electron micrograph of the sintered fibre preform without any hybridization. It is clear from the micrographs that the assorted fibres are randomly oriented and have near-uniform distribution through the preform coupled with a good contact between them. The introduction of fibres, which are in an agglomerated form along with the shots of varying size into the aluminium alloy, does result in a poor distribution of the fibres in the metal matrix. Hence, adequate treatment of the fibre is made with the primary objective of obtaining better properties in the synthesized composites. In **Figure 2(b)** and **Figure 2(c)** is shows the sintered hybrid preform having 20 vol. % of fibre hybridized with 10 vol. % of SiCp and 20 vol. % of SiCp. Here, a near-uniform distribution of the reinforcing SiC particulates was observed between the fibres.

The fabricated preform should have a combination of good inter-connecting porosity, high strength, and high hardness. This is both desired and essential to prevent deformation of the preform during the course of infiltration. The deciding factor governing strength of the preform depends on the kind of binder used during fabrication. **Figure 3** is a scanning electron micrograph of the sintered performs showing the occurrence of bridging between the fibres. The bonding is made possible by the melting of the inorganic binder (aluminium) subsequent to firing. The aluminium binder was observed to be absent at the surface of the fibre.

In **Figure 4** is shown the microstructure of AA319 - alumino-silicate fibre – $SiC_p$ hybrid preform composite that was fabricated using the infiltration process. In Figure 4(a) the infiltrated composite [i.e., 20 vol. % alumino-silicate fiber-10 vol. % $SiC_p$) sample reveals a dense and pore free microstructure with a near homogeneous distribution of the reinforcing phases in the metal matrix. All of the pores were filled with liquid molten metal. The dark or black colour region present between the fibres is the shots, which had varying dimension. The shots must be removed. The cracking of shots coupled with penetration of molten metal into the cracks during infiltration was observed in the microstructure. In Figure 4(b) is shown the microstructure of the hybrid composite (20% alumino-silicate fibre -20% $SiC_p$) having a near uniform distribution of the reinforcing phase. Ultrasonic treatment of the fibres in an aqueous medium much prior to fabrication of the preform does significantly reduce the percentage of shots that can be observed in the microstructure

[Figure 4 (b) compared to Figure 4 (a)]. In Figure 4 (c) is shown microstructure of the interface region of the hybrid preform reinforced aluminium alloy composite when the metal begins to penetrate into the preform. The image analysis results reveal two gradations in structural morphology of the primary aluminium and eutectic silicon phases to be present in the matrix of the chosen aluminium alloy metal matrix. The size of the primary aluminium and eutectic silicon phases was observed to be finer towards the interface region when compared to the outer side. This is because the infiltrated metal diffused out during application of higher pressure and subsequently during solidification. The microstructure of the matrix at the region of the interface was found to be quite similar to microstructure of the matrix of the metal-matrix composite.

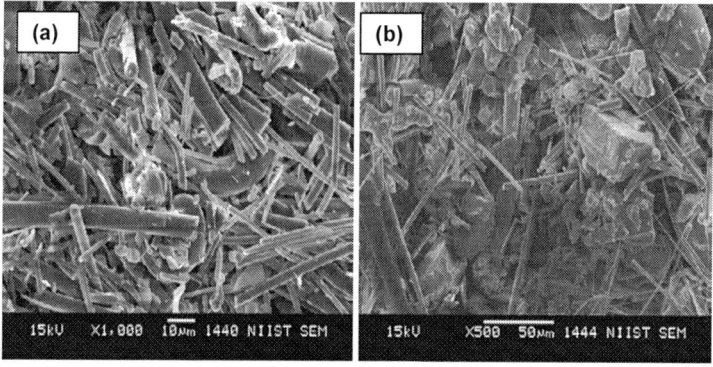

**Figure 2.** Scanning electron micrographs of the:
(a) Sintered alumino-silicate fibre preform without any addition of $SiC_p$.
(b) The sintered hybrid preform having 20 vol.% of fibre hybridized with 10 vol.% of $SiC_p$, and
(c) The sintered hybrid preform having 20 vol. % of fibre hybridized with 20% vol. of $SiC_p$.

**Figure 3.** Scanning electron micrograph of the sintered perform showing bonding between the fibres.

Surface of the alumino-silicate fibre-SiC particulate incorporated in the metal matrix is shown in **Figure 5** and does reveal the occurrence of interfacial reactions. Scanning electron micrograph revealed the occurrence of good interfacial bonding between the alumino-silicate fibre-$SiC_p$ and aluminium alloy metal matrix. This can be safely attributed to a synergism of the following:
(i)   Good infiltration,
(ii)  Formation of $MgAl_2O_4$ spinel at the interface, and
(iii) Presence of aluminium binder used in the hybrid preform.
Earlier studies have shown the presence of a metallic coating or binders on the surface of the reinforcing phase to generally aid in improving the overall wettability of the composite [10]. There is also a possibility of presence of the eutectic silicon on the surface of the fibre.

The X-ray diffraction pattern of both the alumino-silicate fibres and reinforcing SiC particulates extracted from the AA319 composite using sodium hydroxide (NaOH) is shown in **Figure 6**. The XRD analysis reveals the alumino-silicate fibre and the ceramic ($SiC_p$) particulates to react with the aluminium alloy matrix. The peaks of both the $MgAl_2O_4$ and MgO phase is formed by the chemical interaction between magnesium in the matrix and the constituents, i.e., $Al_2O_3$ and $SiO_2$, present in the fibre. Also, magnesium is a powerful surfactant that scavenges oxygen from the melt surface and easily forms the $MgAl_2O_4$ spinel and MgO at the interfaces of the aluminium alloy – alumino-silicate - SiCp. This aids in not only enhancing the wettability but also facilitates effective infiltration of the liquid metal into the perform. During the chemical reactions silicon gets rejected into the matrix and due to its poor solubility in aluminium it tends to normally precipitate and be present as silicon, which tends to segregate to the interface region during solidification of the metal. The $Mg_2Si$ phase forms due to excess addition of magnesium to the aluminium alloy for purpose of enhancing wetting. Therefore, it has to be limited to

4.7 mass% due to the formation of the brittle intermetallic compound $Al_3Mg_2$. The ductile phase $Al_{12}Mg_{17}$ forms as a consequence of a reaction between the aluminium and magnesium, which helps in improving the mechanical properties of the material.

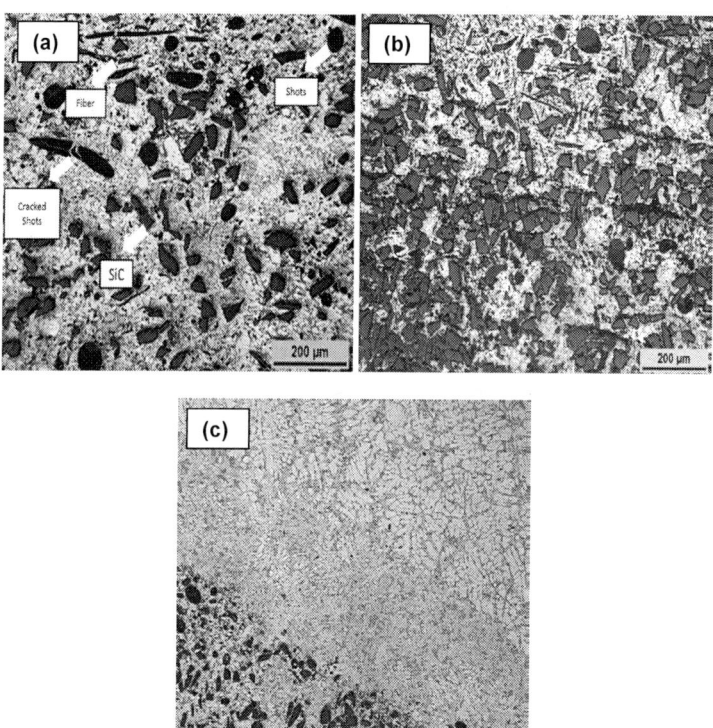

Figure 4.  Scanning electron micrographs showing:
(a)  Microstructure of the AA319 + 20 vol. % alumino-silicatefiber-10% SiC   hybrid preform composite,
(b)  AA319 + 20 vol. % alumino-silicate fiber-20 vol. % SiC hybrid composite, and
(c)  Microstructure of the interface region of the hybrid preform reinforced aluminium alloy composite

**Figure 5.** Scanning electron micrographs of the extracted composite particle showing good interfacial bonding between the alumino-silicate fibres - SiC particle and the aluminium alloy metal matrix

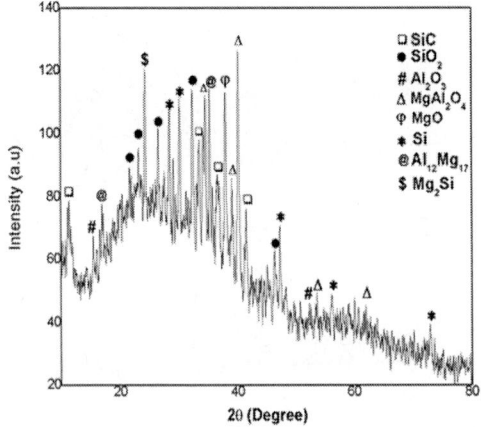

**Figure 6.** XRD pattern of the particle extracted from AA319-20% aluminosilicate fiber-10% SiCp hybrid preform composite using sodium hydroxide (NaOH) solution showing interfacial reaction products

The Brinell hardness values for both aluminium alloy 310 and the composite in both the as- cast and heat treated conditions are summarized in **Table.2**. Presence of the reinforcing phases increases the hardness of the matrix alloy from 119 for the as-cast alloy to 223 BHN for the hybrid composite (20% alumino-silicate fibre -10%$SiC_p$).

Hardness of the processed composite, in both the as-cast and heat treated conditions, is noticeably higher than that of the matrix alloy. The improvement in Brinell hardness due to presence of the reinforcing phase is well over 60 pct.

| Sample Type | Hardness (BHN) (As cast) | Hardness (BHN) (Heat treated) |
|---|---|---|
| Matrix Alloy | 108 | 119 |
| Hybrid composite [AA319 - 20% Aluminosilicate fibre - 10%SiCp) | 166 | 223 |

**Table 2.** Brinell hardness values for both the matrix alloy and hybrid MMC processed using the squeeze infiltration process

## Conclusions

1. Squeeze infiltration of liquid metal into a ceramic preform was successfully carried out with the objective of engineering metal matrix composites having a near-net shape size. This was made possible using hybrid preforms of: (i) 20 vol. % alumino-silicate fiber-10 vol. % SiCp and (ii) 20 vol. % aluminosilicate fiber-20 vol. % SiCp composition.

2. The hybrid preforms were prepared with sufficient porosity and strength by the inorganic salt porogen technique with the addition of alumino-silicate short fibres, silicon carbide particulates (SiCp) and aluminium alloy powder as the binding agent.

3. Aluminium is used as the binder, which helps in lowering the binding temperature of the porous preform while concurrently providing high strength. During infiltration the aluminium binder acts as a metallic coating on the surface of the reinforcement, which contributes to improving the wettability of the composites.

4. The infiltrated composite samples revealed a dense and pore free microstructure having a near homogeneous distribution of both the fibres and particulates in the metal matrix.

5. Ultrasonic treatment of the fibres in an aqueous medium, prior to fabrication of the preform, significantly reduced the percentage of shots. Presence of both the $MgAl_2O_4$ and $MgO$ phases is observed due to a chemical interaction between magnesium in the aluminium alloy matrix and the constituents, i.e., $Al_2O_3$ and $SiO_2$, present in the fibre.

6. For both the as-case and heat treated conditions, hardness of the processed composite was noticeably higher than that of the matrix alloy. The improvement in Brinell hardness due to presence of the reinforcing phase is as high as 60 pct.

## Acknowledgements

The authors are grateful to the *Director* and **Members** of *Materials Science and Technology Division*, CSIR-NIIST, Trivandrum for their support and help. *Mr. M. R. Chandran*, and Mr. *P. Guruswamy* deserve special praise and recognition for their timely assistance with scanning electron microscopy and x-ray diffraction analysis

## References

1. D. Huda, M.A. El Baradie and M.S. Hashmi, *Journal of Materials Processing Technology*, 37 (1993) pp.513-528.

2. T.M. Yue and G. A. Chadwick: *Journal of Materials Processing Technology*, 58, (1996) pp. 302-307.

3. AdemDemir, and A. Necat: *Composites Science and Technology*, Vol. 64 (13-14) (2004), pp 2067-2074.

4. V. Sampath, N. Ramanan and R. Palaninathan: *Materials and Manufacturing Processes*, 21, (2006), pp. 495–505.

5. Paolo Colombo, *Philosophical Transactions R* Soc, A 364(2006), pp.109-124.

6. K.M. SreeManu, V. G. Resmi, M. Brahmakumar, P. Narayanasamy P, T.P.D. Rajan, C. Pavithran and B. C. Pai B C: , *Transactions of Indian Institute of Metals*, 65(6) (2012), pp. 747–751.

7. K. M. SreeManu, V. G. Resmi, M. Brahmakumarm N. Anand, T.P. D. Rajan, C. Pavithran, B. C. Pai and K. Manisekar: *Materials Science Forum*, 710 (2012), pp.371-376.

8. Adam Papworth, and Peter Fox: *Materials Letters*, 29 (1996), pp. 209-213.

9. L.H. Qi, L.Z. Su, J. M. Zhou, J. T.Guan, X.H. Hou, and H. J. Li: *Journal of Alloys and Compounds*, 527 (2012), pp. 10–15.

10. T.P. D. Rajan, R. M. Pillai, and B.C. Pai: *Journal of Materials Science*, 3 (1998) pp. 3491-3503.

# THE MICROSTRUCTURE AND MECHANICAL PROPERTIES OF MAGNETIC SHAPE MEMORY ALLOYS $NiCo_{40+x}Al_{30-x}$ [X=0、3、6、10]

Jia Ju, Feng Xue, Jian Zhou, Jing Bai, Huan Liu

*School of Materials Science and Engineering*
**Southeast University**
Nanjing, Jiangsu 211189, PR China

**Abstract**

Detailed investigations of the microstructure and mechanical properties of $NiCo_{40+x}Al_{30-x}$(X=0、3、6、10) alloys have been performed. The results revealed that the microstructure of specimens were consisted of the primary dendrite and inter-dendritic eutectic structure ($\gamma+\beta$ two phase structure), and martensitic arise in matrix phase emerge good self-accommodation effect. The martensitic transition temperature of sample with high content Co-Al ratio was above room temperature. The mechanical studied by both hardness and compressive test. The $\gamma$ phase appears better ductility than $\beta$ phase. Hence the mechanical properties of the specimens were significantly improved with the volume fraction of $\gamma$ phase. While the volume fraction of $\gamma$ phase from 6% to 63% and the compressive strength from 2759 MPa to 3455 MPa, respectively. In addition, the compressive strain of alloys rise over 45%.

Keywords: Shape memory alloys, Microstructure, Martensitic transformation, Micro-hardness, Mechanical properties, Co-Al ratio

## Introduction

Ferromagnetic shape memory alloys (FSMA) become a very popular research topics in recently, because of broad use in Aerospace, Marine and Land Applications areas [1, 2]. Compared with magnetostrictive alloys (e.g., Terfenol-D), FSMA had larger magnetic field induced strain and rapid response[3]. Additionally, the strain in FSMA can be triggered not only by change in temperature and stress, but also by change in the applied magnetic field [4-6].

Ni-Co-Al system had been developed as a new FSMA recently, which was characterized by good ductility, large magnetic anisotropy and low cost of constituent elements. In Ni-Co-Al systems, the $\beta+\gamma$ tow phase region was more stable with wide range of composition and temperature. Meanwhile, $B_2$-type $\beta$ phase was transition to the $L1_0$-type or modulated $\beta$' martensite, and phase transformation temperature and magnetic transition temperature can be controlled by elements composition [7-20]. As well known, $\gamma$ phase as a second phase in Ni-Co-Al systems could incorporates ductility with $\beta$ phase, but restrain the magnetic and magnetic field induced strain [21-24]. However, if Co-Ni-Al alloys want to get larger magnetic field induced strain and good ductility, the effect of microstructure on the mechanical and magnetic properties must to be clarified. In this paper, the influences of Co-Al ratio in alloys microstructure and mechanical properties in Co-Ni-Al alloys were investigated.

## Experimental

The specimens $NiCo_{40+x}Al_{30-x}$(X=0、3、6、10) were prepared by arc-melting using elements with purity better than 99.99% under argon atmosphere (all specimens composition were listed in Table 1). The alloys were melted four times and the ingot was homogenized at 1475K for 12h, followed by a quenching in ice water. The microstructure was examined by an optical microscopy (OM) and scanning electron microscopy (SEM). In order to determine the phases, X-ray diffraction (XRD) investigations were performed in the Philips PW170 using Cu K$\alpha$ radiation. The phase transformation temperature was determined by STA449 F3 differential scanning calorimetry (DSC) with the cooling and heating rate of 10K/min. The hardness of samples was measured using Vickers hardness tester with load of 20N. The compression test with samples of 3mm×3mm×5mm were performed on CMT5105 machine at the cross-head displacement speed of 0.05mm/min.

Table 1. Composition, microstructure volume fraction and grain size of test specimens

| Alloy designation | Alloy composition (At. %) | Volume fraction of γ phase ($V_\gamma$/%) | Grain size ($d$/μm) | Co-Al ratio |
|---|---|---|---|---|
| A20 | Ni-50%Co-20%Al | 63 | 26 | 2.50:1 |
| A24 | Ni-47%Co-23%Al | 38 | 35 | 2.04:1 |
| A27 | Ni-44%Co-26%Al | 11 | 48 | 1.69:1 |
| A30 | Ni-40%Co-30%Al | 6 | 83 | 1.33:1 |

## Results and Discussion

### Microstructures and phase crystal structures

Figure 1 show the XRD patterns taken from specimens A20 to A30. Obviously, samples were detected as martensite and γ phase mixture. A minor peak indexed to be (111) for FCC. structure martensite phase emerges at 44° in Figure 1. The martensite with a $L1_0$ structure and the tetragonal distortion c/a is 0.86. In addition, it also can be seen that besides (111) peak, (220) and (200) γ phase peak were exhibited in all samples. But (111) peak becomes weak and (220)、(200) peak increase along with Co-Al ratio rise. That means alloys had more precipitants of γ phase with Co-Al ratio rise (the volume fraction of γ phase percentage show in Table 2). On the other hand, martensite phase transformation temperature was rather sensitive to the change elements composition. In alloy A30 martensite peak was distinct. But with Co-Al ratio arise, the peak of martensite decrease and martensite translate into γ phase. The phase structures determined by XRD had a good agreement with the following result of optical observation.

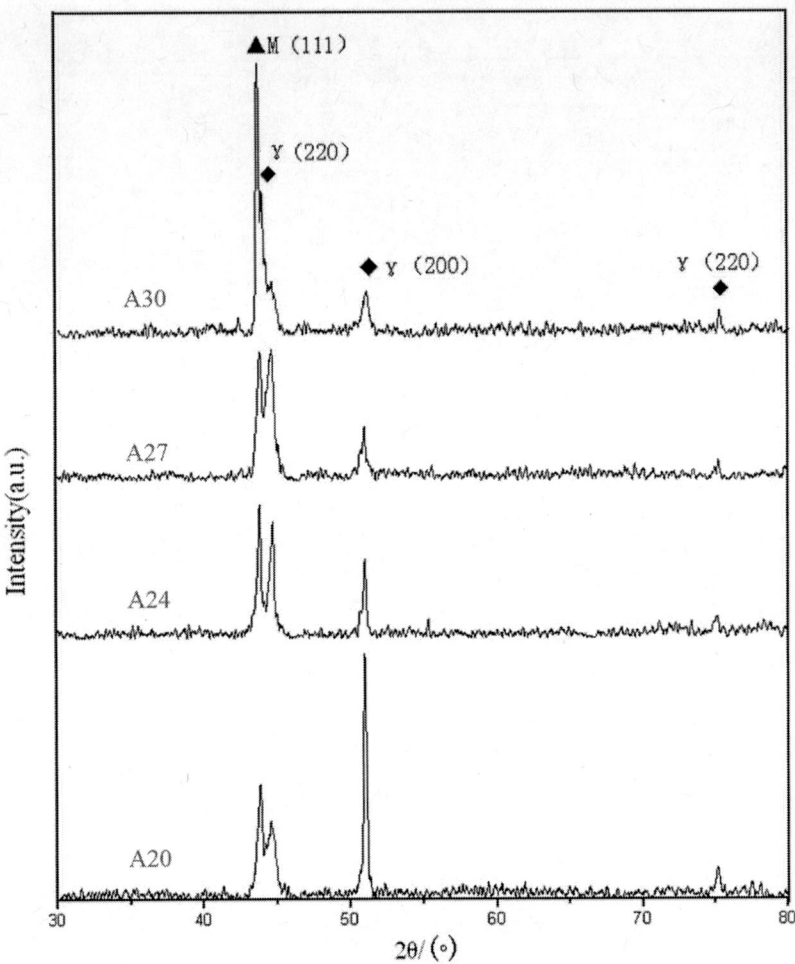

**Figure 1.** X-ray diffraction patterns of the specimens

The typical microstructures of specimens with β+γ tow phase structure were shown in Figure 2. All the alloys consisted of the primary dendrite and inter-dendritic eutectic structure. From Figure 2A to Figure 2D, Co rich γ phase becomes coarse along with Co content rise. The γ phase arise in grain boundary first as seen in Figure 2A and Figure 2B, then relatively large γ phase precipitates in the matrix phase (Figure 2C and Figure 2D) and γ phase arise in the grain boundary becomes coarse from Figure 2A to Figure 2D.

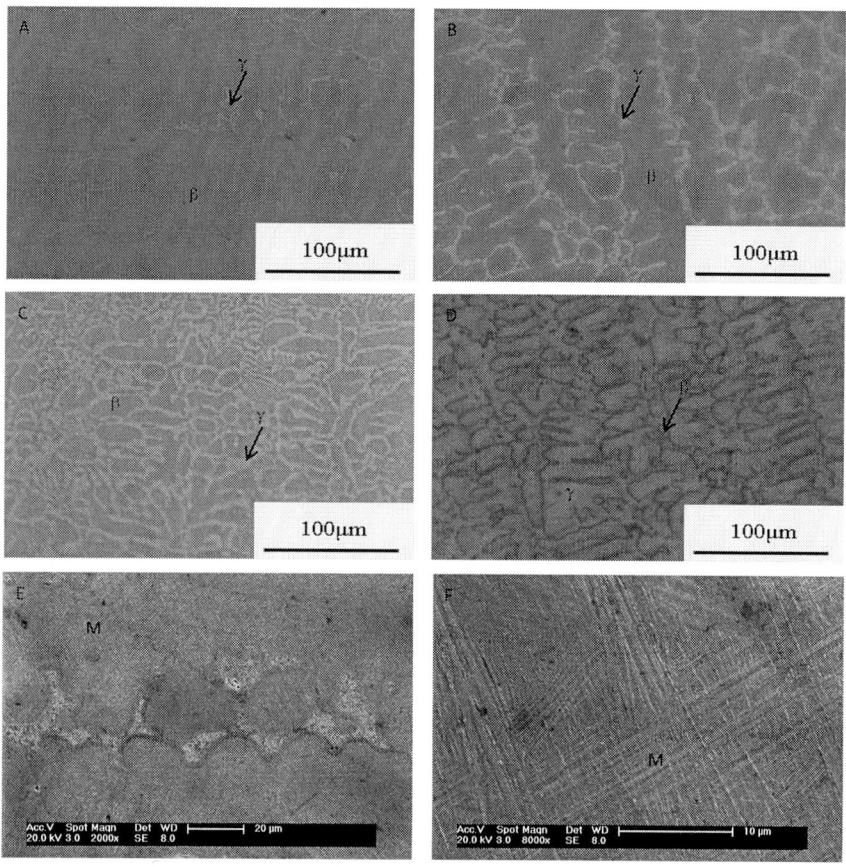

Figure 2. Optical micrographs and SEM image of alloy A20 to A30
(A) Alloy A30; (B) alloy A27; (C) alloy A24; ( ) alloy A20; (E) SEM image of alloy A27; (F) SEM image of alloy A27

Significantly, martensite were observed in matrix phase (shown in Fig.2E and Fig.2F) and appeared self-accommodation. The appearance of the martensite confirmed that martensite transformation temperature of alloys undergone above room temperature. It was also in good agreement with the following result of DSC analysis.

The matrix grain size measurement conducted by line-cross method. The grain size and volume fraction of γ phase were shown in Table 2 and Figure 3. Apparently, the matrix grain size decreases with Co-Al ratio increased. The grain size was about 83μm, 48μm, 35μm and 26μm while the Co-Al ratio was 1.33:1, 1.69:1, 2.04:1 and 2.5:1, respectively.

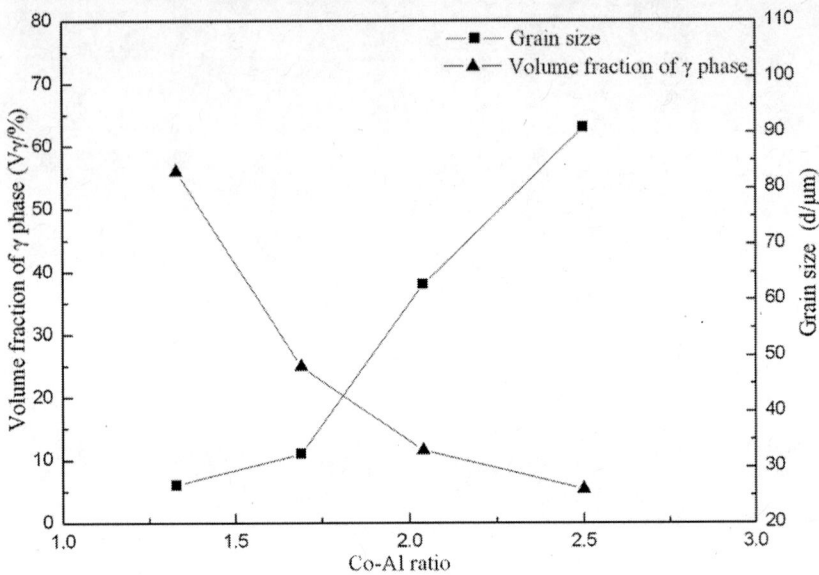

Figure 3:   The volume fraction of γ phase and grain size in samples

**Martensitic transitions**

Figure 4 present the DSC curves of alloy A20 to A30. The temperature of forward martensitic transition ($M_s$) were listed in Table 2.

The value of $M_s$ decreases with increasing of Co-Al ratio and the thermal hysteresis in phase transition was just about 40K. The less thermal hysteresis was one single of the thermo elastic martensite. Indeed, the martensitic transition temperature of alloys can be plotted as a function of the average electron concentration and atomic radius. However, Co element valence electrons and atomic radius was 27 and 1.67Å; Al element valence electrons and atomic radius were 13 and 1.82Å. Obviously, the valence electrons density of Co is higher than Al. Perhaps, this was the reason of $M_s$ decrease along with Co-Al ratio rise.

Table 2. Phase transition temperature and thermal hysteresis of samples (/K)

| Alloy | $M_s$ | $M_f$ | $A_s$ | $A_f$ | $\triangle T = A_f - M_s$ |
|---|---|---|---|---|---|
| A20 | 296 | 291 | 309 | 335 | 39 |
| A24 | 306 | 299 | 324 | 343 | 37 |
| A27 | 330 | 313 | 347 | 370 | 40 |
| A30 | 344 | 338 | 356 | 391 | 47 |

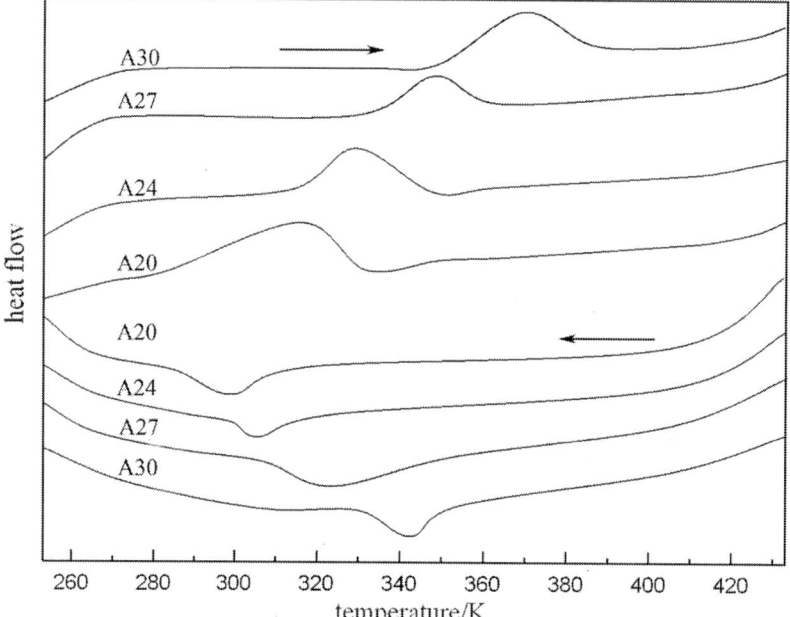

Figure 4. DSC analyses show influence of sample elements composition on martensitic transformation temperature in alloy A20 to A30

## Mechanical properties

The results of hardness test of samples were presented in Figure 5. The hardness of alloy is changing from 498 HV up to 678 HV and this increase was governed by the increase of Co-Al ratio. In the other words, the ductility of alloy enhances along with γ phase increase, because of the γ phase had more ductility than matrix phase. This was confirmed with micro-hardness tester shown in Figure 6.

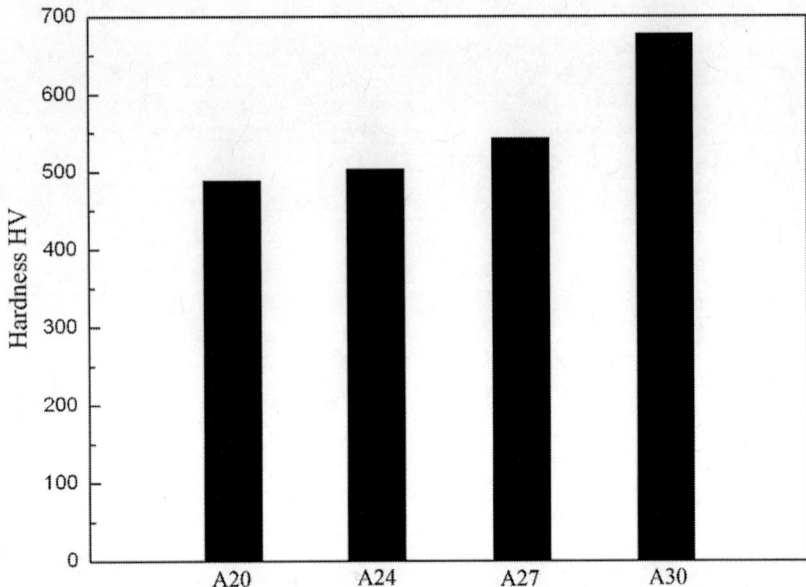

**Figure 5. Result of hardness measurements of investigated samples**

For all the specimens, compression test had been carried out at room temperature. The compression strain-stress curves were shown in Figure 7 and the result of ultimate compression test were shown in Table III. Clearly, it was found that alloys contained more γ phase exhibit excellent compressive ductility at room temperature.

Form Table 3 and Figure7, one can see that alloy A20 has the best mechanical properties, compressive strength approach 3455 MPa and the compressive strain was more than 45%. By contrast, samples content lower Co-content and higher Al-content had poor ductility such as alloy A30. Obviously, the mechanical properties were strongly improved by the increased volume fraction of γ phase.

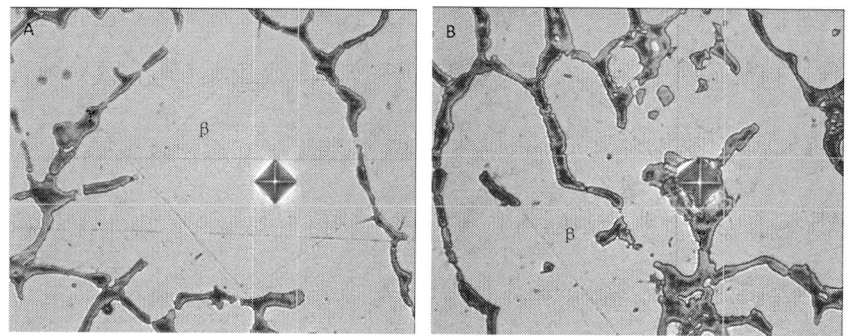

Figure 6. Micro-hardness tester of β and γ phase in alloy A24
(A) Micro-hardness of β phase: 713 HV;
(B) Micro-hardness of γ phase: 454 HV)

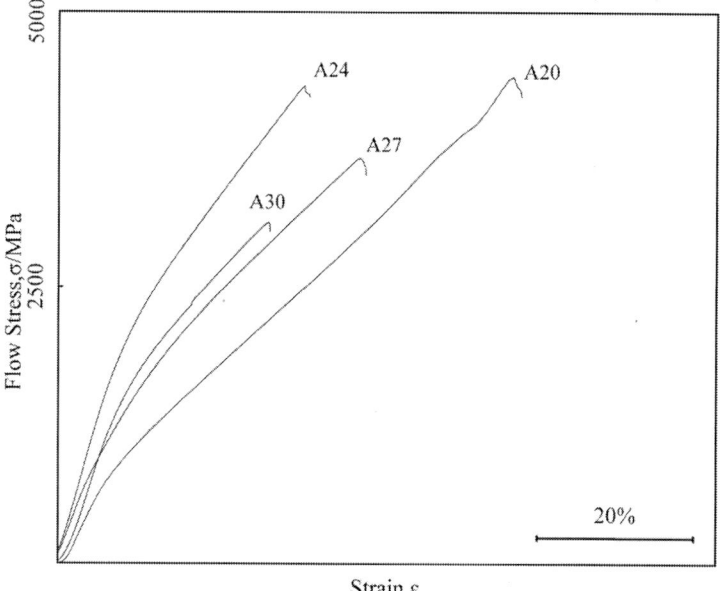

Figure 7. Stress-strain curves from compression test for alloys A20 to A30.

Table 3.    The result of compressive strength and compressive strain in alloy A20 to A30

| Alloy | Compressive strength (MPa) | Compressive strain (%) |
|---|---|---|
| A30 | 2759 | 21 |
| A27 | 3090 | 32 |
| A24 | 3168 | 27 |
| A20 | 3455 | 48 |

**Conclusions**

The microstructure and mechanical properties in Ni-Co-Al alloys had been investigated. The results obtained were as follows:

1. The structure observations revealed that the specimens consisted of the primary dendrite and inter-dendritic eutectic structure ($\gamma+\beta$ phase structure) in all alloys. And along with Co-Al ratio rise FCC. structure $\gamma$ phase arise in grain boundary and becomes coarse. Furthermore, form SEM image martensite can be seen in matrix phase and shown good self-accommodation effect. Additionally, the grain size was 26μm in alloy A20, but with Co-Al ratio increase, the grain size grows up to 83 μm in alloy A30.

2. In the $NiCo_{40+x}Al_{30-x}$(X=0、3、6、10) alloys, with influence of the average electron concentration and atomic radius, Ms temperature was decreases with addition of Co-Al ratio. In alloys A24, A27 and A30, martensitic transition from the matrix phase to the $L1_0$ martensitic phase were over the room temperature.

3. The result of hardness test was A30>A27>A24>A20. Compared with $\beta$ phase, $\gamma$ phase has more ductility. Hence the alloy A30 had good hardness because of the specimen had more $\gamma$ phase arise in grain boundary when Co-Al ratio increases.

4. The mechanical properties of the specimens were significantly improved with the volume fraction of $\gamma$ phase. With all samples the volume fractions of $\gamma$ phase were from 6% to 63% and the compressive strength was from 2759MPa added to 3455 MPa, respectively. In addition, the compressive strain of alloys rise over 45%.

# References

1. Ganor Y, Shilo D, Zarrouati N, James RD. Ferromagnetic shape memory flapper. *Sensors and Actuators- PHYSICAL* 2009;150 (2):277-9.

2. Ullakko K. Magnetically controlled shape memory alloys: A new class of actuator materials. *Journal of Materials Engineering and Performance* 1996; 5 (3), 405-9.

3. FUNAYAMA T, SAKAI I. Mn Substitution Effect on Magnetostriction Temperature-Dependence in TB0.3DY0.7FE2. *Applied Physics Letters,* 1992, 61(1), 114-5.

4. Sozinov A, Ullakko K. Crystal structures and magnetic anisotropy properties of Ni-Mn-Ga martensitic phases with giant magnetic-field-induced strain. *IEEE Transactions on Magnetics* 2002, 38(5), 2814-6.

5. O'Handley RC. Model for strain and magnetization in magnetic shape-memory alloys. Journal of Applied Physics 1998, 83(6), 3263-70.

6. Ullakko K, Kantner C, Kokorin VV. Large magnetic-field-induced strains in Ni2MnGa single crystals. *Applied Physics Letters* 1996, 69 (13), 1966-8.

7. Li J, Li J. Effects of solidification rate and temperature gradient on microstructure and crystal orientation of Co32Ni40Al28 alloys. *Materials Letters* 2012;68:40-3.

8. Scheerbaum N, Kraus R, Liu J, Skrotzki W, Schultz L, Gutfleisch O. Reproducibility of martensitic transformation and phase constitution in Ni‐Co‐Al. *Intermetallics* 2012;20(1):55-62.

9. Li JZ, Huang B, Li JG. Growth of large crystallites of Co37Ni34Al29 ferromagnetic shape memory alloys under super-high temperature gradient directional solidification. *Journal of Crystal Growth* 2011;317(1):110-4.

10. Maziarz W. Structure changes of Co‐Ni‐Al ferromagnetic shape memory alloys after vacuum annealing and hot rolling. *Journal of Alloy and Compounds,* 2008; 448(1-2):223-6.

11. Liu J, Li JG. Microstructure, shape memory effect and mechanical properties of rapidly solidified Co - Ni - Al magnetic shape memory alloys. *Materials Science and Engineering: A* 2007; 454 - 455:423-32.

12. Liu J, Zheng HX, Li JG. Effect of solidification rate on microstructure and crystal orientation of ferromagnetic shape memory alloys Co - Ni - Al. *Materials Science and Engineering: A* 2006, 438 - 440(04):1061-4.

13. Liu J, Li JG. Magnetic force microscopy observations of Co - Ni - Ga and Co - Ni - Al alloys with two-phase structures. *Scripta Materialia* 2006;55(9):755-8.

14. Valiullin AI, Kositsin SV, Kositsina II, Kataeva NV, Zavalishin VA. Study of ferromagnetic Co - Ni - Al alloys with thermoelastic L10 martensite. *Materials Science and Engineering: A* 2006, pp. 1041-4.

15. Wang HY, Liu ZH, Wang YG, Duan XF, Wu GH. Microstructure of martensitic phase in the Co39Ni33Al28 shape memory alloys revealed by transmission electron microscopy. *Journal of Alloys and Compounds* 2005, 400(1 - 2), pp.145-9.

16. Hiraga K, Ohsuna T, Sun W, Sugiyama K. The structural characteristics of Al - Co - Ni decagonal quasicrystals and crystalline approximants. *Journal of Alloys and Compounds* 2002, 342(1 - 2), pp. 110-4.

17. Oikawa K, Wulff L, Iijima T, Gejima F, Ohmori T, Fujita A, Fukamichi K, Kainuma R, Ishida K. Promising ferromagnetic Ni-Co-Al shape memory alloy system. *Applied Physics Letters,* 2001, 79(20), pp. 3290-2.

18. Cervellino A, Haibach T, Steurer W. Modeling atomic surfaces for the Al - Ni - Co basic decagonal phase. *Materials Science and Engineering: A* 2000, 294 - 296, pp. 276-8.

19. Sato TJ, Hirano T, Tsai AP. Single-crystal growth of the decagonal Al - Ni - Co quasicrystal. *Journal of Crystal Growth* 1998, 191(3), pp. 545-52.

20. Kainuma R, Ise M, Jia CC, Ohtani H, Ishida K. Phase equilibria and microstructural control in the Ni-Co-Al system. *Intermetallics* 1996; Vol. 4, Supplement 1, pp. S151-8.

21. Khandelwal A, Sharma VK, Sharath Chandra LS, Arora P, Chattopadhyay MK, Roy, S.B. The magnetic properties across the martensitic transition in the Co38Ni34Al28 alloy. *Journal of Magnetic Materials* 2012, 324(5), pp. 729-34.

22. Morito H, Oikawa K, Fujita A, Fukamichi K, Kainuma R, Ishida K. Large magnetic-field-induced strain in Co – Ni – Al single-variant ferromagnetic shape memory alloy. *Scripta Materialia*, 2010, 63(4), pp. 379-82.

23. Chatterjee S, Thakur M, Giri S, Majumdar S, Deb AK, De SK. Transport, magnetic and structural investigations of Co – Ni – Al shape memory alloy. *Journal of Alloys and Compounds* 2008, 456(1 – 2), pp. 96-100.

24. Tanaka Y, Oikawa K, Sutou Y, Omori T, Kainuma R, Ishida K. Martensitic transition and superelasticity of Co – Ni – Al ferromagnetic shape memory alloys with $\beta + \gamma$ two-phase structure. *Materials Science and Engineering: A* 2006, 438 – 440, pp. 1054-60.

# METAL MATRIX COMPOSITES DIRECTIONALLY SOLIDIFIED

Alicia Esther Ares [1,2], Carlos Enrique Schvezov [1,2]

[1] Materials Institute of Misiones, IMAM (CONICET-UNaM)
University of Misiones
1552 Azara Street, Posadas, Misiones, 3300 Argentina.

[2] Member of CIC of the National Research Council (CONICET) of Argentina.

## Abstract

The present work is focused on the study of the effect of the directional solidification on the SiC particles distribution inside the zinc-aluminum matrix during the columnar – to – equiaxed (CET) phenomenon. The following parameters were measured: cooling rates, temperature gradients, interphase velocities on the solidification microstructure of the MMCs. The following tests were carried out on the composite: optical microscopy, EDS, Vickers microhardness (HV) and Electrochemical Impedance Spectroscopy (EIS) technique. The primary dendrite arm spacing in composite is smaller than in the matrix alloy. The HV values of composites directionally solidified are higher than the HV values of alloys in all structures (columnar, CET and equiaxed). The results also indicate that the corrosion resistance of Zn-Al-SiC composites has demonstrated the improvement in comparison to ZA alloys in 3wt.% NaCl solutions.

*Keywords:* Metal Matrix Composites, Zinc-Aluminum, Microhardness, Corrosion resistance

## Introduction

The mechanical properties of zinc-aluminum (ZA) alloys are generally considered satisfactory from room temperature up to about 100 °C. Above this temperature, the applications are diminished due to the decrease in the creep resistance [1-7]. It is also known that metal matrix composite (MMCs) usually exhibits improved creep properties compared to the unreinforced matrix. With respect to the addition of elements, copper additions are more effective than silicon additions in improving the mechanical properties of the ZA alloys; however silicon improves the wear resistance of the alloy as well as its dimensional stability [7].

The presence of SiC particles in the ZA alloys increases the value of the elastic modulus and the hardness. On the other hand, there is an improvement with the addition of alumina particles or glass fiber. Adding ceramic particles improves the strength and stiffness, but reduces the ductility. The wear resistance of the alloy increases with the volume fraction of the reinforcement [4-6].

In previous research, the directional solidification of ZA-27 and ZA-50 alloy matrix reinforced with SiC particles was analyzed [8-9].

In the present research, because there is no sufficient experimental data about the columnar-to- equiaxed transition phenomenon on metal matrix composites (MMC) directionally solidified in the literature, in the present research the CET was analyzed in ZA-8, ZA12 and ZA-27 alloys reinforced with SiC particles. Although, it mechanical and electrochemical properties were observed.

## Experimental Procedure

The casting alloys (Zn-8wt.%Al, Zn-12wt.Al and Zn-27wt.%Al) were prepared from Zn 99.998 wt.% and aluminum 99.999 wt.%, and composites by using the same alloys as matrix and adding SiC (p) in different quantities in volume percent. The alloys and composites were directionally solidified and K-type thermocouples, protected with ceramic shields, were longitudinally arranged using to measure thermal profiles during the growth [10-12]. The samples were melted in ceramic molds. The solidification time was approximately 100 s. After solidification, the samples were cut and polished with emery paper up to 1000 grit and then using 1 um diamond paste. Samples were etched for approximately 10-20 s, at room temperature (20-25 °C) in a solution of chromic acid (i. e., 50 g $Cr_2O_3$; 4 g $Na_2SO_4$ in 100 ml of water [13]). The position of the CET was determined by observation under Arcano® and Olympus® optical microscopes. The microstructure was analized using SEM and an image analyzer. The compositions of the samples which are considered in this report are shown in Table I.

Microhardness measurements were performed at room temperature with a Buehler® microhardness tester. Loads of 200 gf were used. The measurements were performed under ASTM E 384-89 standard using a pressing time of 15 s. Both bulk alloys and directionally solidified samples were used.

The primary dendrite spacings were measured in the cross sections of the other half of the samples near the thermocouple positions using an image analyzer (Neophot). Each section was mounted, polished and etched and the spacing was determined by the number of intercepts along a straight line.

For the electrochemical tests, samples of 20 mm in length of each zone and for each concentration were prepared as test electrodes, polished with sandpaper (from SiC #80 until #1200) and washed with distilled water and dried by natural flow of air. All the electrochemical tests were conducted in 3wt% NaCl solution at room temperature using an IM6d Zahner®- Elektrik potentiostat coupled to a frequency analyzer system.

### Results and Discussion

The liquidus and solidus temperatures for each alloy were determined using the differential thermal analysis system, NETZSCH STA 449 C with calibrated cells with pure elements. The DTA measurements were done with a charge of 200 mg of pre-melted samples in alumina crucibles. The heating and cooling velocities were 10 °C/min using Argon atmosphere. The cycles were repeated several times to check for reproducibility of the results, the liquidus, $T_L$, and solidus, $T_S$, temperatures were taken from the heating and cooling curves as usual [14-16]. The values obtained are presented in Table I.

The columnar-to-equiaxed transition (CET) was obtained in a number of twenty seven experiments. A typical macrostructure of the CET is shown in Figure 1 (a) for ZA-12 - 15vol%SiC composite. In Figures 1(b) to 1(d) it is possible to appreciate the resultant microstructures in the columnar, CET and equiaxed zones of macrostructure in Figure 1 (a).

As reported elsewhere for other alloys the transition is not sharp showing a zone where some equiaxed grains co-exist with columnar grains [17-23]. The size of the transition zone is in the order of up to ten millimeter between the minimum position of the CET (CET Min.) and the maximum position of the CET (CET Max.). Also, no effect of the set of the thermocouples in the CET was observed; either acting as nucleating sites or changing the solidification structure.

Table I.   Liquidus ($T_L$) and solidus ($T_S$) temperatures for each ZA alloy (metal matrix) and composite.

| Composite | $T_L$ (K) | $T_S$ (K) |
|---|---|---|
| Zn-8wt.%Al (ZA8) | 698 | 655 |
| Zn-12wt.%Al (ZA12) | 705 | 655 |
| Zn-27wt.%Al (ZA27) | 778.2 | 709.7 |
| ZA8-8vol%SiC | 622.5 | 598.2 |
| ZA8-15vol%SiC | 642.1 | 648.1 |
| ZA12-8vol%SiC | 646.3 | 622.7 |
| ZA12-15vol%SiC | 672.4 | 623.8 |
| ZA27-8vol%SiC | 718.3 | 704.9 |
| ZA27-15vol%SiC | 725.4 | 702.1 |

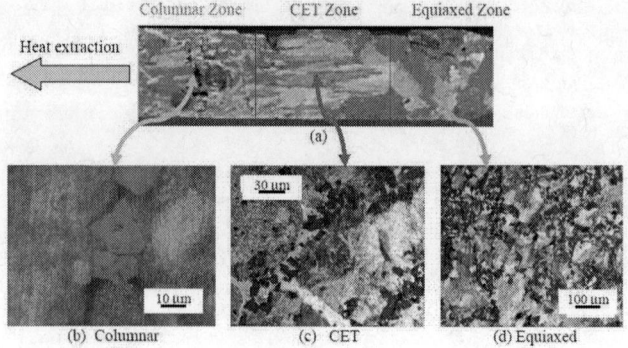

**Figure 1.** (a) Macrostructure of in MMCs (ZA12 -15vol%SiC) showing the columnar-to- equiaxed transition.
(b) to (d) Microstructures of different zones showing the dendrite growth and SiC particles.

The time dependent temperatures measured by the thermocouples in each position of the samples with a CET, is shown in Figure 2 for ZA12-8vol%SiC. In the figure the thermocouple T1 is at the lowest position and the first to reach the solidification front and T5 is at the highest position. In all the curves it is possible to identify a period corresponding to the cooling of the melt, a second period of solidification and the final period of cooling of the solid to ambient temperature.

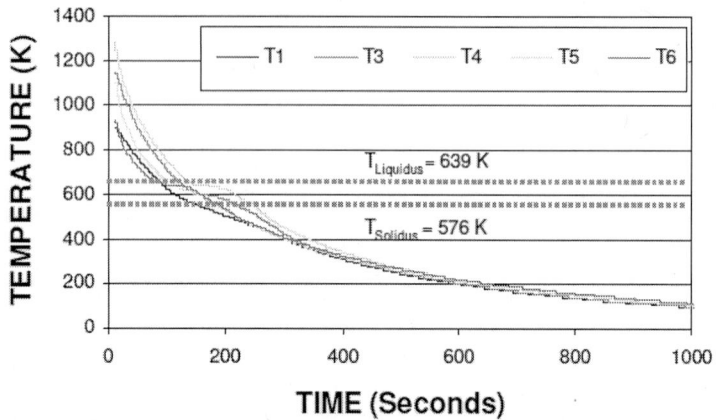

**Figure 2.** Temperature versus time curves. ZA12-8vol%SiC.

Table II lists the location of the CET from the bottom of the sample, which is in the range of CET MIN. to CET MAX. (cm). Comparing the cooling rates with the distances, which correspond to the length of the columnar zone, it is observed that increasing the velocity increases the length of the columnar grains. The temperature versus time curves also show that the temperature evolution depends on the structure being formed. During columnar solidification, the temperature decreases steadily and monotonically, on the contrary in the equiaxed region, during the transition, there is a recalescence which increases the temperature from a minimum; the level of recalescence for each experiment and is listed in Table II as REC. (°C).

The position of the liquidus and solidus fronts were detected at each thermocouple position by the start and end of solidification. Both temperatures were determined by the change of slope of the cooling curves occurring at event. The velocities of the solidification fronts were calculated by dividing the separation distance between consecutive thermocouples, and the time taken by each front to pass both thermocouple positions. These values are defined as the velocities of the liquidus front, $V_L$, and solidus front, $V_S$. The velocity of the liquidus front at the moment of the CET is called critical liquidus velocity, $V_{LC}$. The values of the critical velocity for each experiment are listed in Table I.

The temperature gradients, G, were calculated for each pair of neighbor thermocouples as the temperature difference between the thermocouple readings divided by the separation distance between thermocouples [18]. The values of gradients are plotted in Figure 3 for ZA12- 8vol%SiC composite. In the figure it is observed that from the beginning of solidification, the gradients decreases with time. The minimum value corresponds to the position of the columnar- to-equiaxed transition (CET). The gradients shown for experiment in Figure 3 reach negative value at this point, which is –1.87 °C/cm. As it is shown in Table II, the gradients determined in most experiments are negative. As was reported for alloys, this negative value is an indication of a reversal in the temperatures profiles ahead of the interface, which could be associated to the recalescence due to a massive nucleation of equiaxed grains, and previously reported and discussed for different alloy systems [17-23]. The fact that in some cases the position of the thermocouples is not located at the precise position where the CET occurs, may prevent detection of the negative gradients which is believe to occur in all cases. Nevertheless, the values always reach a minimum value at this position as in the case of alloys.

In order to analyze the effect of adding ceramic particles to the ZA alloys in the columnar-to- equiaxed transition, it is noted that after comparing the results for similar cooling rates in alloys with and without particles it is possible to appreciate that in the case of the alloys with particles increase the length of the columnar zone with respect to the alloys without particles for similar cooling rate. This can be seen by comparing the results in line 10 with line 1 in Table I for ZA8 without and with SiC particles. Similarly, by comparing the results in line 7 with line 22 for the case of ZA27 alloy and composite, where for a cooling rate of 1.6 K/s the minimum length of columnar zone is 24 mm and 21 mm, respectively, and for a cooling rate of 2.7 K/s the minimum length of columnar zone is 37 mm and 49 mm, respectively. The effect that be remarked in this case is an increase on the length of the position of the transition zone.

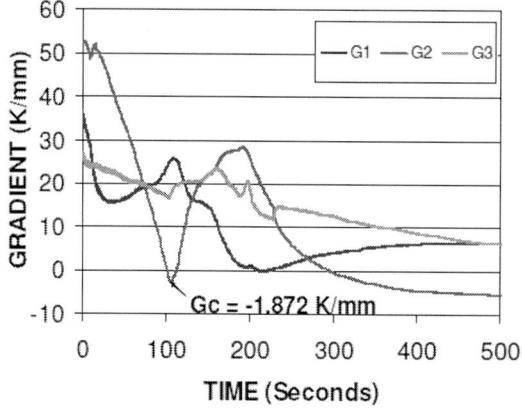

**Figure 3.**   Gradients versus time. ZA12-8vol%SiC composite.

119

Figure 4 shows the variation in the primary dendrite arm spacing, $\lambda_1$, with growth velocity in directionally solidified samples of ZA27-16vol.%SiC composite under different temperature gradients. As expected, the primary dendrite arm spacing decreases with an increase in growth rate and / or an increases in temperature gradient ( black +).
Experimental results on primary dendrite arm spacing in an Zn-27wt.%Al alloy are also superimposed in the same figure ( white <>). It is interesting to note that for a given solidification condition, $\lambda_1$ in composite is smaller than that in the base alloy. However, this effect is more pronounced in the low-growth-rate regime and diminishes at higher growth rates.

**Table II.** Liquid cooling rates ($\dot{T}_L$) and solid ($\dot{T}_s$), critical liquidus interphase velocity (V$_{LC}$), CET minimum position (CET $_{MIN}$) and maximum (CET $_{MAX}$) and critical gradient (G$_C$) obtained from the temperature versus time curves.

| # | ALLOY/ COMPOSITE | $\dot{T}_L$ (°C/s) | $\dot{T}_s$ (°C/s) | V$_{LC}$ (mm/s) | CET$_{MIN}$ (mm) | CET$_{MAX}$ (mm) | G$_c$ (K/mm) |
|---|---|---|---|---|---|---|---|
| 1 | Zn-8wt.%Al (ZA8) | 1.6 | 1.2 | 1.1 | 15 | 27 | -1.25 |
| 2 | Zn-8wt.%Al (ZA8) | 2.2 | 1.7 | 1.7 | 22 | 35 | -0.95 |
| 3 | Zn-8wt.%Al (ZA8) | 2.7 | 1.9 | 1.9 | 33 | 67 | -0.52 |
| 4 | Zn-12wt.%Al (ZA12) | 1.6 | 1.3 | 1.0 | 18 | 29 | -1.11 |
| 5 | Zn-12wt.%Al (ZA12) | 2.2 | 1.6 | 1.5 | 24 | 40 | 0.09 |
| 6 | Zn-12wt.%Al (ZA12) | 2.7 | 1.8 | 2.0 | 39 | 71 | -1.63 |
| 7 | Zn-27wt.%Al (ZA27) | 1.6 | 1.2 | 0.9 | 21 | 33 | -2.01 |
| 8 | Zn-27wt.%Al (ZA27) | 2.2 | 1.6 | 1.7 | 29 | 42 | -0.55 |
| 9 | Zn-27wt.%Al (ZA27) | 2.7 | 2.0 | 1.9 | 37 | 50 | -0.36 |
| 10 | ZA8-8vol%SiC | 1.6 | 1.4 | 1.2 | 19 | 31 | -1.96 |
| 11 | ZA8-8vol%SiC | 2.2 | 1.7 | 1.5 | 27 | 42 | -0.77 |
| 12 | ZA8-8vol%SiC | 2.7 | 2.3 | 1.8 | 37 | 69 | -098 |
| 13 | ZA8-16vol%SiC | 1.6 | 1.2 | 1.1 | 18 | 28 | -2.07 |
| 14 | ZA8-16vol%SiC | 2.2 | 1.7 | 1.7 | 30 | 45 | -2.56 |
| 15 | ZA8-16vol%SiC | 2.7 | 1.9 | 1.9 | 42 | 73 | -2.11 |
| 16 | ZA12-8vol%SiC | 1.6 | 1.1 | 0.9 | 23 | 37 | -1.61 |
| 17 | ZA12-8vol%SiC | 2.2 | 1.6 | 1.7 | 36 | 52 | -1.87 |
| 18 | ZA12-8vol%SiC | 2.7 | 1.9 | 1.9 | 55 | 79 | -0.78 |
| 19 | ZA12-16vol%SiC | 1.6 | 1.4 | 1.2 | 17 | 34 | -0.37 |
| 20 | ZA12-16vol%SiC | 2.2 | 1.7 | 1.5 | 31 | 56 | -0.41 |
| 21 | ZA12-16vol%SiC | 2.7 | 2.3 | 1.8 | 52 | 77 | -0.63 |
| 22 | ZA27-8vol%SiC | 1.6 | 1.2 | 1.1 | 24 | 46 | -2.19 |
| 23 | ZA27-8vol%SiC | 2.2 | 1.6 | 1.7 | 35 | 53 | -1.57 |
| 24 | ZA27-8vol%SiC | 2.7 | 2.0 | 1.9 | 49 | 66 | -2.26 |
| 25 | ZA27-16vol%SiC | 1.6 | 1.3 | 1.0 | 31 | 44 | -1.34 |
| 26 | ZA27-16vol%SiC | 2.2 | 1.6 | 1.5 | 46 | 61 | -1.68 |
| 27 | ZA27-16vol%SiC | 2.7 | 1.8 | 2.0 | 58 | 82 | -1.46 |

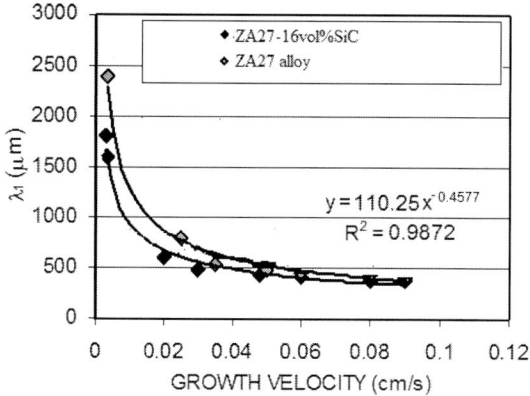

**Figure 4.** Primary dendrite spacing, $\lambda_1$, versus growth velocity for ZA27-16vol%SiC composite and ZA27 alloy matrix.

In addition, the Vickers microhardness, HV, for the same composite is plotted as a function of the distance from the bottom of the sample (black ● in Figure 5 (a)). It is observed that HV increase steadily from the columnar to equiaxed region. Similar results were obtained for HV in ZA27 alloy (white <>). The values of HV in composite are bigger than in alloy and increase steadily in the columnar zone with a peak in the CET zone, then the HV decrease in the equiaxed region (white !:: in Figure 5 (a)).

The Vickers microhardness, HV, for composite was plotted as a function of the width of the sample (black ● in Figure 5 (b)). In is possible to observe that the HV is smaller in the center of the sample and increase on both sides.

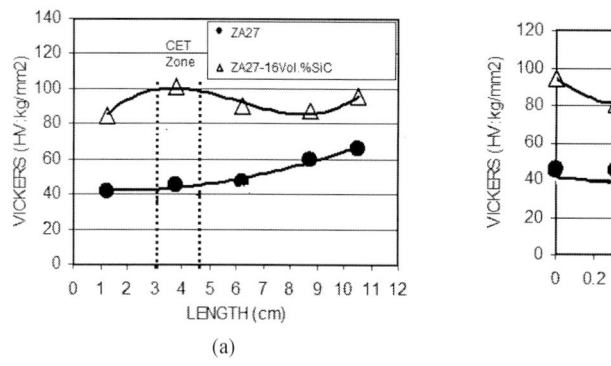
(a)

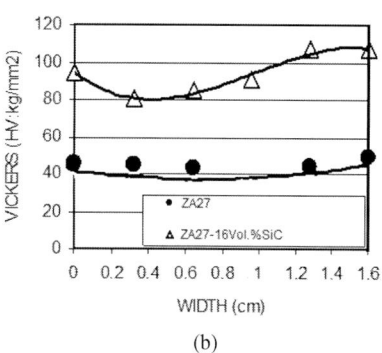
(b)

**Figure 5.** Vickers microhardnees versus (a) length and (b) width of ZA27-16vol%SiC composite and ZA27alloy matrix.

Likewise, from EIS analysis and comparing the charge resistance transference, $R_{ct}$, values was possible to establish that the ZA27 alloy with different structures is less resistant to corrosion and its susceptibility to corrosion is dependent of the structure, see Table III. The ZA27-16%vol.CSi composite has the highest corrosion resistance. Also, discriminating by type of composite material, the MMCs with SiC are more corrosion resistant than those MMCs prepared with alumina particles.

**Table III.** Principal parameters obtained from EIS analysis: $R\Omega$ is the ohmic solution resistance, $C_{dl}$ is the double layer capacity and $R_{ct}$ is the charge transfer resistance.

| ALLOY/COMPOSITE | $R\Omega$ ($\Omega.cm^2$) | $C_{dl}$ ($F/cm^2$) | $R_{ct}$ ($\Omega.cm^2$) |
|---|---|---|---|
| Zn-27-8vol.%CSi | 14.5 | $3.65e^{-5}$ | 942 |
| ZA27-16vol.%SiC | 13.5 | $7.95\ e^{-5}$ | 1320 |
| Zn-27-16vol.%Al$_2$O$_3$ | 13 | $6.94e^{-5}$ | 152 |
| Zn-27-8vol.%Al$_2$O$_3$ | 11.8 | $4.22e^{-5}$ | 140 |
| Zn-27wt.%Al (Columnar) | 8.9 | $4.9e^{-5}$ | 95 |
| Zn-27wt.%Al (CET) | 9.5 | $1e^{-4}$ | 27 |
| Zn-27wt.%Al (Equiaxed) | 8.6 | $5.8e^{-5}$ | 87 |

## Conclusions

From the results of the previous sections the main conclusions of this investigation on the columnar to equiaxed transition in composites are:

1. It was possible to obtain values of the thermal parameters associated with the CET of ZA8, ZA12 and ZA27 composites and compare them with those obtained for ZA alloys.

2. In ZA composites the transition occurs in a similar form to the case of ZA alloys, that is, in a zone rather than in a sharp plane, where both columnar and equiaxed grains in the melt co-exist.

3. In the conditions of these experiments, the columnar length for composites increases respect of the columnar length for ZA alloys.

4. The primary dendrite arm spacing, decreases with an increase in growth rate and / or an increase in temperature gradient, like occurs with alloys.

5. The primary dendrite arm spacing, ☐1 in composite is smaller than in the matrix alloy. However, this effect is more pronounced in the low-growth-rate regime and diminishes at higher growth rates.

6. The HV values of composites directionally solidified are higher than the HV values of alloys in all structures (columnar, CET and equiaxed).

7. The results also indicate that the corrosion resistance of Zn-Al-SiC composites has demonstrated the improvement in comparison to ZA alloys in 3wt.% NaCl solutions.

## Acknowledgements

This work was supported by PICT-2011-1378 of the National Agency for Promotion of Science and Technology.Thanks are due to the Argentinean Research Council (CONICET) for the financial support.

## References

[1] M. Rappaz, Modelling of microstructure formation in solidification process. International Materials Reviews 1989; 34 (3): 93-123.

[2] J.E. Spittle, Columnar to equiaxed grain transition in as solidified alloys. International Materials Reviews 2006; 51 (4): 247-269.

[3] M. Sahoo, L.V. Whiting, Foundry characteristics of sand cast Zn-Al alloys. AFS Transactions 1984; 92: 861-870.

[4] M. Sahoo, L.V. Whiting, W.G. Whited, Control of underside shrinkage in Zinc-aluminum foundry alloys by the addition of trace elements. AFS Transactions 1985; 93: 475-480.

[5] E. Gervais, H. Levert, Bess M. The development of a family of zinc base foundry alloys. AFS Transactions 1980; 88: 183-194.

[6] D. Delneuville, Tribological behaviour of Zn-Al alloys (ZA27) compared with bronze when used as a bearing material with high load and at very low speed. Wear 1985; 105: 283-292.

[7] R. Auras, C. E. Schvezov, Wear behavior, microstructure, and dimensional stability of as-cast zinc-aluminum/SiC (metal matrix composites) alloys. Metallurgical and Materials Transactions A 2004; 35: 1579-1590.

[8] A.E. Ares, C.E. Schvezov, International Foundry Research/Giessereiforschung 60 (2008) 2-9.

[9] A.E. Ares, C.E. Schvezov, Columnar-to-equiaxed transition in metal matrix composites, 142nd Annual Meeting and Exhibition: Linking Science and Technology for Global Solutions, TMS 2013; San Antonio, TX; United States; 3 March 2013 through 7 March 2013.

[10] Ares AE, Caram R, Schvezov CE. Solidification Structures and Properties of Zinc- Aluminium /SiC (MMC) Alloys, Solidification Processing of Metal Matrix Composites. In: Proceedings of Solidification Processing of Metal Matrix Composites – Rohatgi Honorary Symposium. 2006 TMS Annual Meeting. San Antonio, Texas, USA, March, 2006. p. 183-193.

[11] Ares AE, Schvezov CE. Solidification parameters during the columnar-to-equiaxed transition in Lead-Tin alloys. Metallurgical and Materials Transactions A 2000; 31: 1611-1625.

[12] Ares AE, Schvezov CE. Influence of solidification thermal parameters on the columnar-to- equiaxed transition of Aluminum-Zinc and Zinc-Aluminum alloys. Metallurgical and Materials Transactions A 2007; 38:1485-1499.

[13] G.F. Vander Voort, Metallography Principles and Practice. New York: ASM International, 2000).

[14]  RF. Speyer, Thermal Analysis of Materials, New York: Marcel Dekker, 1994.

[15]  YT Zhu, JH. Devletian, Application of Differential Thermal Analysis to Solid-Solid Transitions in Phase Diagram. Journal of Phase Equilibria 1994; 15: 37-41.

[16]  W.J. Moffatt, Handbook of Binary Phase Diagrams. New York: General Electric Company Corporate Research and Development Technology Marketing Operation, 1984.

[17]  A.E. Ares, S.F.Gueijman, C.E. Schvezov, Semi-empirical modeling for columnar and equiaxed growth of alloys. Journal of Crystal Growth 2002; 241: 235-240.

[18]  A.E. Ares, S.F. Gueijman, R. Caram, C.E. Schvezov, Analysis of solidification parameters during solidification of lead and aluminum base alloys. Journal of Crystal Growth 2005; 275: 319-325.

[19]  R.B. Mahapatra, F. Weinberg, The columnar to equiaxed transition in Tin-Lead alloys. Metallurgical and Materials Transaction B 1987; 18: 425-432.

[20]  I. Ziv, F.Weinberg, The columnar-to-equiaxed transition in Al 3 Pct Cu. Metallurgical and Materials Transaction B 1989; 20: 731-734.

[21]  T.M. Pollock, W.H. Murphy, The breakdown of single-crystal solidification in high refractory nickel-base alloys. Metallurgical and Materials Transactions A 1996; 27: 1081-1094.

[22]  Ch. A. . Gandin, Experimental study of the transition from constrained to unconstrained growth during directional solidification. ISIJ International 2000; 40 (10): 971-979.

[23]  Ch. A. Gandin, From constrained to unconstrained growth during directional solidification. Acta Materialia 2000; 48: 2483-2501.

# MECHANICAL AND MATERIAL PROPERTY EVALUATION

Advanced Composites for Aerospace, Marine, and Land Applications
Edited by: Tomoko Sano, T.S. Srivatsan, and Michael W. Peretti
TMS (The Minerals, Metals & Materials Society), 2014

# A NEW CLASS OF METAL NANOCOMPOSITES WITH SUPERIOR MECHANICAL PROPERTIES: UNUSUAL THERMAL EXPANSION IN NbTi-NANOWIRES / NiTi-MATRIX COMPOSITE

Shijie Hao[1], Lishan Cui[1*], Shan Wang[1], Daqiang Jiang[1], Cun Yu[1], Dennis E. Brown[2], Yang Ren[3*]

[1]Department of Materials Science and engineering
China University of Petroleum
Beijing 102249, China

[2]Department of Physics
Northern Illinois University
De Kalb, Illinois 60115, USA

[3]X-ray Science Division
Argonne National Laboratory
Argonne, IL 60439, USA

*Corresponding Authors: lishancui63@126.com; ren@aps.anl.gov

## Abstract

An *in-situ* composite composed of NbTi nanowires, which are well dispersed and aligned in amorphous/nanocrystalline NiTi matrix, exhibits an unusual non-linear thermal expansion behavior during heating from 25 °C to 600 °C. The evolution of the structure of NiTi matrix and the axial thermal expansion property of NbTi nanowires during heating process were studied by *in situ* synchrotron X-ray diffraction. The collective effects of the intrinsic thermal expansion properties of NiTi matrix and NbTi nanowires, and the coupling between nanowires and matrix during heating were responsible for the unusual thermal expansion behavior. The unusual non-linear thermal expansion property may have potential applications, such as thermal actuator, sensor and fastener materials.

Keywords: Metal nanocomposite, Thermal expansion, High-energy X-ray diffraction

## Introduction

Thermal expansion which related to the thermal vibration of atoms and the variation of mean interatomic distance during heating is one of the most important physical properties of a material [1,2]. Generally, the metallic materials exhibit a positive thermal expansion behavior on heating due to the increase of mean interatomic distance with the temperature increase [1-3]. Metallic materials with extreme or unusual thermal expansion behavior are of interest from both technological and fundamental standpoints. Examples include materials with zero thermal expansion, maximum thermal expansion, and negative thermal expansion [4-8]. Heretofore, however, it is still not easy for material scientists to design and fabricate the metallic materials with extreme or unusual thermal expansion properties. In this study, we report an unusual non-linear thermal expansion behavior in a NbTi nanowire *in-situ* composite with amorphous/nanocrystalline NiTi matrix during heating from 25 °C to 600 °C, which is significantly different from that of general metallic materials and metal-matrix composites with ceramic fibers [3,9-12]. The unusual non-linear thermal expansion property may make the composite very useful in a range of applications such as thermal actuator, thermal sensor and thermal fastener [13,14].

## Results and discussion

An alloy ingot with a composition of $Ni_{38}Ti_{42}NbTi_{20}$ (at. %) was fabricated by vacuum induction melting, then it was hot forged at 850 °C into rods with 16 mm in diameter. The hot-forged rod was further hot drawn at 750 °C into wire with 1 mm in diameter and then cold drawn at room temperature into thin wire of 0.58 mm in diameter. The test sample was cut from the thin cold-drawn wire. *In situ* synchrotron X-ray diffraction measurement was performed at the 11-ID-C beamline of the advanced photon source. High energy X-rays with a beam size of 0.6 mm × 0.6 mm and wavelength of 0.10798 Å were used to obtain two-dimensional (2-D) diffraction pattern in transmission geometry. The heating rate is 10 °C /min and the diffraction patterns were recorded every 5 °C.

Transmission electron microscope (TEM) and scanning electron microscope (SEM) investigations shown in Figures 1(a) and 1(b) reveal that the NbTi nanowires (NbTi solid solution containing Ti 16.4 and Ni 3.1 (at. %)) ranging from hundreds of nanometers to hundreds of microns in length are well dispersed and aligned in NiTi matrix along the wire axial direction. The average diameter of the nanowires is 78 nm. The 2-D diffraction pattern shown in Figure 1(e) indicate that the [110] direction of NbTi nanowires is well oriented along the wire axial direction and the NiTi matrix contains amorphous phase, B2 phase and B19' phase [15]. The TEM dark-field image (Figure 1(c)) and the high-resolution TEM image of the NiTi matrix (Figure 1(d)) further indicate the nanocrystals (B2-NiTi and B19'-NiTi) dispersed in the amorphous phase. The volume fractions of NbTi nanowires and NiTi matrices are about 25 % and 75 %, respectively.

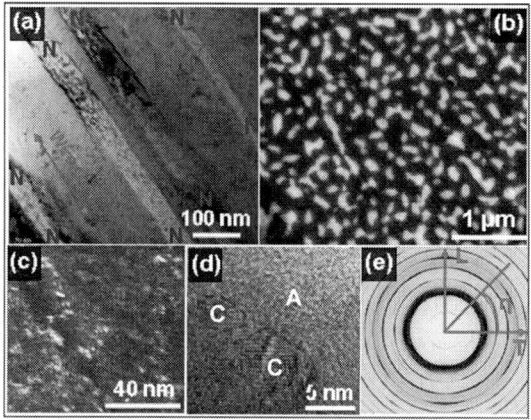

Figure 1. (a) TEM micrograph of longitudinal-section of the composite (NbTiTi nanowires marked by "N"),
(b) SEM micrograph of transverse-section of the composite (bright points: the cross sections of NbTiTi nanowires),
(c) TEM dark-field image of the NiTi matrix,
(d) High-resolution TEM image of the NiTi matrix (nanocrysals marked by "C" and amorphous marked by "A"),
(e) 2-D X-ray diffraction pattern of the composite. The longitudinal L and transverse T directions and the azimuthal angle η are denoted.

Figure 2 shows the thermal expansion curve of the cold-drawn composite wire. The corresponding data of the ingot specimen is also included in Figure 2. It can be seen that the cold-drawn composite wire exhibits an unusual non-linear thermal expansion behavior during heating in comparison with the ingot specimen. According to the magnitude of thermal expansion coefficient, the thermal expansion curve of the cold-drawn composite wire can be roughly divided into four sections: A-B, B-C, C-D and D-E (Figure. 2). The temperature ranges and the thermal expansion coefficients of the four sections are summarized in Table. 1.

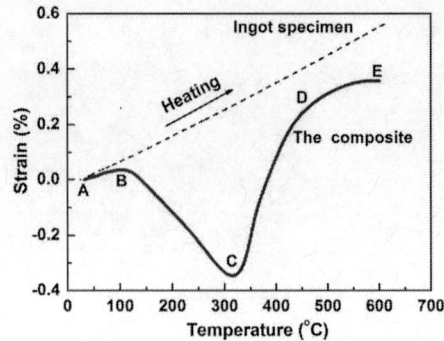

**Figure 2.** Thermal expansion curves of the cold-drawn composite wire and the ingot specimen.

**Table 1.** The temperature ranges and the corresponding thermal expansion coefficients of the cold-drawn composite wire, the NbTi nanowires and the ingot specimen.

| Section number | First (A-B) | Second (B-C) | Third (C-D) | Forth (D-E) |
|---|---|---|---|---|
| Temperature range | 25→110 °C | 110→315 °C | 315→500 °C | 500→700 °C |
| The composite: α | $5.4 \times 10^{-6}$ | $-2.04 \times 10^{-5}$ | $6.2 \times 10^{-5}$ | $-5.6 \times 10^{-6}$ |
| The nanowires: α | $6.1 \times 10^{-6}$ | $-1.7 \times 10^{-7}$ | $7.1 \times 10^{-5}$ | $9.6 \times 10^{-6}$ |
| The as-cast specimen: α | $8.4 \times 10^{-6}$ | | | |

α: Thermal expansion coefficient

To reveal the origin of unusual thermal expansion behavior, the evolution of the structure of amorphous/nanocrystalline NiTi matrix and the variation of the axial thermal expansion property of NbTi nanowires during heating were studied by *in situ* synchrotron X-ray diffraction. Figure 3(a) shows the plot of the *d*-spacing of $(110)_{NbTi}$ versus the azimuthal angle "η" for the cold-drawn specimen before heating. The longitudinal **L** and transverse **T** angles are labeled. This result indicates that a significant internal compressive strain on the NbTi nanowires along the wire axial direction was introduced during the preparation process of the composite. The evolution of the lattice strain for $(110)_{NbTi}$ (i.e., the axial lattice strain of the nanowires) as a function of temperature is shown in Figure 3(b). According to the expansion rate, the plot of the axial lattice strain of NbTi nanowires vs. temperature is also can be roughly divided into four sections, which correspond well to the four sections on the thermal expansion curve of the specimen (Figure 2). The expansion coefficients of NbTi nanowires on the four sections are also summarized in Table. 1. The evolutions of the diffraction peaks for $(110)_{B2-NiTi}$, $(211)_{B2-NiTi}$, $(110)_{NbTi}$ and $(220)_{NbTi}$ as a function of temperature are shown in Figure 3(c). The intensities of the diffraction peaks of B19'-NiTi phase are too weak to be clearly observed in Figure 3(c), and thus the evolution of the diffraction peak for $(001)_{B19'-NiTi}$ with the temperature is shown in the inset in Figure 3(d). The plot of the diffraction peak area of $(001)_{B19'-NiTi}$ versus the temperature is shown in Figure 3(d). It can be seen that, in the temperature range from 110 °C to 315 °C, the B19'-NiTi phase in matrix reversely transformed to B2-NiTi. When the temperature is above 315 °C, the amorphous phase begin to crystallize, which causes a significant increase in the intensity of the diffraction peaks of B2-NiTi phase (Figure 3(c)) and the appearance of endothermic (crystallization) peak on the differential scanning calorimeter curve (Figure 4(a)).

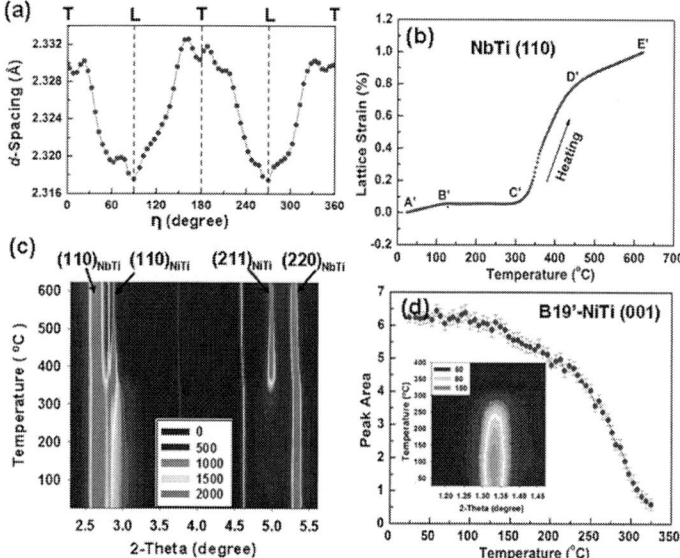

Figure 3. (a) The plot of the $d$-spacing for $(110)_{NbTi}$ versus the azimuthal angle "η" for the cold-drawn specimen before heating,
(b) The plot of the lattice strain for $(110)_{NbTi}$ perpendicular to the loading direction versus the temperature,
(c) The evolution of the diffraction peaks for $(110)_{B2-NiTi}$, $(211)_{B2-NiTi}$, $(110)_{NbTi}$ and $(220)_{NbTi}$ as a function of temperature,
(d) The plot of the diffraction peak area of the $(001)_{B19'-NiTi}$ versus the temperature, Insert of Figure 1(d) is the evolution of the diffraction peaks for B19'-NiTi (001) with the temperature.

Based on the above experimental results, we offer the following rationalization for the unusual thermal expansion behavior. For the first section of the thermal expansion curve (A-B in Figure 2), no change in the structures of the NiTi matrix and the NbTi nanowires was observed and the thermal expansion coefficients of the composite and the nanowires are consistent with that of the ingot specimen. Thus, this positive thermal expansion mainly originates from the initial thermal expansion of matrix and nanowires with increasing temperature. The negative thermal expansion on the second section (B-C in Figure 2) can be attributed to the reverse transformation of the B19'-NiTi phase formed during the preparation of the composite (i.e., B19'-NiTi→B2-NiTi). From Figure 3(d), it can be seen that the deformed martensitic NiTi phase reversely transformed to austenitic NiTi phase in this section, which caused a contraction of the matrix of the composite along the axial direction during heating [16]. Because of the contraction of the matrix, the NbTi nanowires were subjected to an increasing compressive stress from the matrix during heating, which leads to the NbTi nanowires exhibiting a slightly negative expansion in this section (B'-C' in Figure 3(b)). From Figure 3(c) and Figure 4(a), it can be observed that the amorphous/nanocrystalline NiTi matrix occurred crystallization on the third section (C-D in Figure 2). Generally speaking, amorphous crystallization leads to an increase in density and a decrease in volume [17,18]. But, the composite exhibits an unusual rapid positive expansion during the crystallization of amorphous/nanocrystalline NiTi matrix. From Figure 3(b), we note that the NbTi nanowires exhibit a rapid elongation along the axial direction during the crystallization of NiTi matrix. This rapid elongation originates from the compressive strain on the NbTi nanowires, which was introduced during the preparation of the composite (Figure 3(a)) and the contraction of the composite from 110 °C to 315 °C (B-C in Figure 2), was rapidly released during the crystallization of NiTi matrix. Thus the rapid elongation of NbTi nanowires causes the unusual positive thermal expansion of the composite. On the fourth section (D-E in Figure 2), we note that the expansion coefficient of the composite is slightly less than that of the ingot specimen. The plot of the full width at half maximum (FWHM) for $(211)_{B2-NiTi}$ diffraction peaks versus temperature (Figure 4(b)) indicates the width of the diffraction peaks from NiTi matrix decreases with the temperature increase. Because the internal stress in the matrix is very week after crystallization, the rapid decrease in the width of diffraction peaks is mainly attributed to the grain growth of the matrix [19]. The contribution of non-linear crystal lattice vibration in the grain boundary component is apparent because of random atomic arrangement [20,21]. Thus, the rapid grain growth and the quick decrease of the volume fraction of grain boundaries result in the less expansion coefficient of the composite in comparison with that of the ingot specimen.

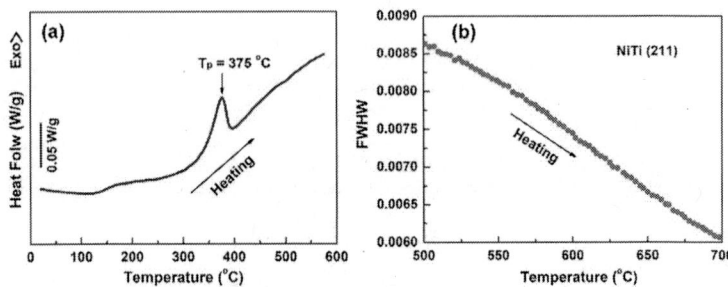

**Figure 4.**  (a)  The differential scanning calorimeter (DSC) curve obtained from the cold-drawn specimen on heating process,
(b)  The plot of the full width at half maximum (FWHM) for $(211)_{NiTi}$ versus the temperature from 450 °C to 600 °C.

## Conclusion

In summary, an unusual non-linear thermal expansion behavior was found in a NbTi nanowire *in-situ* composite with amorphous/nanocrystalline NiTi matrix. With the temperature increase from 25 °C to 600 °C, the amorphous/nanocrystalline NiTi matrix underwent reverse transformation of B19'-NiTi phase, crystallization of amorphous phase and grain growth, and the NbTi nanowires were subjected to increasing compressive strain followed by a rapid relaxation of the compressive strain. As result of the intrinsic thermal expansion (thermal vibration of atoms) of nanowires and matrix and the interaction between matrix and nanowires during heating, this material exhibits the unusual non-linear thermal expansion behavior. This investigation provides an important example and microstructural information for the design and preparation of the metallic materials with extreme or unusual thermal and mechanical property.

## Acknowledgements

This work was supported by the key National Natural Science Foundation of China (NSFC) (51231008), the Science Foundation of China University of Petroleum, Beijing (No.2462013YJRC005) and the Key Project of Chinese Ministry of Education (313055). Use of the Advanced Photon Source was supported by the US Department of Energy, Office of Science, and Office of Basic Energy Science under Contract No. DE-AC02-06CH11357.

## References

1. S. I. Novikova, *Thermal Expansion of Solids* (Nauka, Moscow, 1974).

2. G. Grimvall, *Thermophysical Properties of Materials* (Elsevier, New York, 1999).

3. F.C. Nix and D. McNair, "The Thermal Expansion of Pure Metals: Copper, Gold, Aluminum, Nickel, and Iron", *Physical Review*, 60 (1941), 597-605.

4. Y. Tanji, "Thermal Expansion Coefficient and Spontaneous Volume Magnetostriction of Fe-Ni (fcc) Alloys", *Journal of the Physical Society of Japan*, 31 (1971), 1366-1373.

5. R. Kainuma et al., "Invar-type effect induced by cold-rolling deformation in shape memory alloys", *Applied Physics Letters*, 80 (2002), 4348.

6. T. Saito et al., "Multifunctional Alloys Obtained via a Dislocation-Free Plastic Deformation Mechanism", *Science*, 300 (2003), 464.

7. T.A. Mary et al., "Negative Thermal Expansion from 0.3 K to 1050 K in $ZrW_2O_8$", *Science*, 272 (1996), 90-92.

8. K. Fukamichi et al., "Invar-type new ferromagnetic amorphous Fe-B alloys", *Solid State Communications*, 23 (1977), 955-958.

9. R. Arpon et al., "Thermal expansion be-havior of aluminum/SiCcomposites with bimodal particle distributions", *Acta Materialia*, 51 (2003), 3145-3152.

10. G. Li and W.D. Fei, "Abnormal thermal expansion behavior of aluminum borate whisker reinforced aluminum composite containing $Fe_3O_4$ particles", *Materials Chemistry and Physics*, 99 (2006), 34–38.

11. Y.L. Shen, "Thermal expansion of metal-ceramic composites: a three-dimensional analysis", *Materials Science and Engineering A*, 252 (1998), 269-275.

12. R.U. Vaidya and K.K. Chawla, "Thermal expansion of metal-matrix composites", *Composites Science and Technology*, 50 (1994), 13-22.

13. O. Sigmund and S. Torquato, "Design of materials with extreme thermal expansion using a three-phase topology optimization method", *Journal of the Mechanics and Physics of Solids*, 45 (1997), 1037-1067.

14. O. Sigmund and S. Torquato, "Composites with extremal thermal expansion coefficients", *Applied Physics Letters*, 69 (1996), 3203.

15. K. Otsuka and X. Ren, "Physical metallurgy of Ti–Ni-based shape memory alloys", *Progress in Materials Science*, 50 (2005), 511-678.

16. Y.N. Liu, Y. Liu and J. Van. Humbeeck, "Two-way shape memory effect developed by martensite deformation in NiTi", *Acta Materialia*, 47 (1998), 199-209.

17. J.E. Shelby, "Thermal expansion of amorphous metals", *Journal of Non-Crystalline Solids*, 34 (1979), 111-119.

18. W.K. Njoroge, H.W. Woltgens and M. Wuttig, "Density changes upon crystallization of Ge2Sb2.04Te4.74 films", *Journal of Vacuum Science and Technology A*, 20 (2002), 230-233.

19. C.E. Krill and R. Birringer, "Estimating grain-size distributions in nanocrystalline materials from X-ray diffraction profile analysis", *Philosophical Magazine A*, 77 (1998), 621-640.

20. K. Lu and M.L. Sui, "Thermal expansion behaviors in nanocrystalline materials with a wide grain size range", *Acta Metallurgica et Materialia*, 43 (1995), 3325-3332.

21. H.J. Klam, H. Hahn and Gleiter, "The thermal expansion of grain boundaries", Acta Materialia, 35 (1987), 2101-2104.

# DATA-FUSION NDE FOR PROGRESSIVE DAMAGE QUANTIFICATION IN COMPOSITES

Jefferson Cuadra[1], Prashanth A. Vanniamparambil[1], Kavan Hazeli[1], Ivan Bartoli[2] and Antonios Kontsos[1,*]

[1] Mechanical Engineering & Mechanics Department
Drexel University
Philadelphia, PA, USA

[2] Civil, Architectural and Environmental Engineering Department
Drexel University
Philadelphia, PA, USA

## Abstract

The objective of this article is to present a progressive damage quantification framework for fiber reinforced polymer composites (FRPC) that have widespread use in aerospace and wind-turbine applications. To this aim, a novel optico-acoustic nondestructive evaluation (NDE) setup is presented based on integration of Digital Image Correlation (DIC), Acoustic Emission (AE), and Infrared Thermography (IRT). DIC and IRT full-field strain and temperature maps reveal early development of structural hot spots, associated with locations where inelastic strains accumulate, damage initiates, and final fracture occurs in both tensile and fatigue experiments. Damage quantification is further related to: (i) energy dissipation, (ii) residual stiffness, (iii) average both temporal and spatial temperature variations, and (iv) AE features in time and frequency domains. The extracted NDE parameters suggest three characteristic stages of fatigue life that can be used to construct appropriate models for reliable remaining life-predictions.

**Keywords:** fiber-reinforced composites; nondestructive evaluation; damage; fatigue

## Introduction

The increasing use of composite materials, and in particular fiber reinforced polymer composites (FRPC), in a wide variety of applications including aerospace, energy and others, is due to their excellent mechanical properties, such as high specific stiffness and strength, resistance to environmental effects and good fatigue life [1-5]. However, although composites have been under development for more than two decades they still suffer from unpredictable damage and failure processes mostly because of their inherent heterogeneity and anisotropy compared e.g. to metals. In fact, composite degradation is (as expected) progressive as subcritical damage accumulates throughout the material, however is also less visually evident compared e.g. to striations, slip bands and other surface effects observed in common metallic materials [6], which results in severe difficulties to accurately monitor and quantify their damage. In this context, nondestructive evaluation (NDE) offers perhaps the only monitoring approach to *in situ* quantify developing damage in composite materials, as demonstrated in this article.

In fiber-reinforced composites, failure is controlled by the activation and development of numerous micromechanisms both interlaminate (i.e. delamination) and intralaminar (e.g. matrix cracking, fiber pullout, etc) [7-10]. Similarly, complex loading conditions, such as fatigue and impact, further add to the complications of reliably characterizing the mechanical and damage behavior of composites. Additional complexities exist due to variability reported in manufacturing and machining operations that often result in non-visible defects, as well as local damage. Consequently, failure in composites is highly likely to initiate in the vicinity of such weakened areas the exact spatial distribution of which is in most cases not known a priori [4]. Furthermore, the simultaneous activation of two or more of such failure mechanisms and their unpredictable evolution with applied loading create the need for the development of effective strategies capable to detect early damage pre-cursors, as well as subsequent damage initiation, evolution and accumulation [9]. In this direction, NDE methods have provided substantial information towards understanding the complex damage behavior in advanced materials such as fiber reinforced composites. Reported NDE methods in both monotonic and cyclic loading conditions include digital image correlation (DIC) [11-14], acoustic techniques [15-19] and spectroscopy, as well as electrical potential/resistance methods [20-22]. In addition to NDE methods, other monitoring approaches have been often validated and used as inputs to damage quantification and modeling by relying on the use of extensometers, strain gages, and other contact metrological devices [9, 23-26].

Although reported efforts have succeeded in characterizing the mechanical behavior and damage in specific types of composites, an effective approach which could universally indicate damage initiation and progression and could have the potential to be transitioned to *on board/on site* full scale testing of composite materials is yet to be established [10]. To address this need, a data fusion approach is presented in this article based on the intelligent combination of three complementary (NDE) techniques including DIC, acoustic emission (AE) and infrared thermography (IRT). In fact, the integration of perhaps the only acoustic technique (AE) with actual *in situ* damage monitoring capabilities with the most reliable full field optical metrology method (DIC), which provides microstrain accurate deformation measurements, and a complementary full field technique (IRT) that can cross-validate the other two methods, provides a novel NDE approach with the potential to determine damage precursors, track developing damage and provide inputs to life estimation models, as presented in this article.

## Experimental Setup and Material Information

The NDE setup used to obtain the results presented in this article is shown in Fig.1. The three NDE systems were integrated to a servohydraulic mechanical frame in order to both correlate with the applied loading profile, as well as to achieve adaptive sampling (needed for long fatigue monitoring tests).

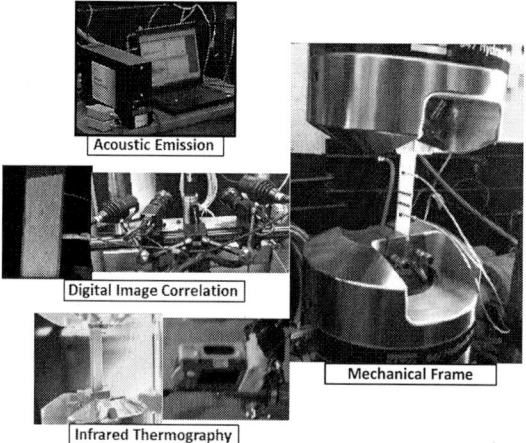

**Figure 1:** Experimental setup including three NDE methods and a mechanical loading frame.

The composites tested were glass fiber reinforced epoxy laminates consisting of E-Glass fibers and epoxy resin. The prepeg system had a stacking sequence of [0/90/90/0] with a fiber volume fraction of 54%. The FRPC specimens were tested under both tension and tension-tension fatigue. The tensile tests were conducted in displacement control at a rate of 2mm/min based on ASTM D3039. The fatigue tests were designed based on the ultimate tensile strength (UTS) measured in the tensile tests and they were carried out at a frequency rate of 3Hz with a stress ratio ($R$-value) of 0.1, at three stress levels of 52%, 62% and 67% of UTS. The AE data acquisition system consisted of a 4-channel AEWin DiSP system (MISTRAS) with two piezoelectric transducers (PICO) and preamplifiers with a uniform gain of 40dB. The two sensors were mounted at the top and bottom of the specimen as shown in Fig 1, using a cyanoacrylate adhesive. The piezoelectric transducers have an operating frequency range of 200-750 kHz with a peak frequency at 500kHz. The received signals were amplified using 2/4/6-AST pre-amplifiers while a threshold of 60dB. Full field deformation measurements were provided by a GOM ARAMIS 3D 5 megapixel DIC system which for the cameras and resolution used has a peak image acquisition rate of 30 frames per second (fps). A random speckle pattern was used on the surface of interest in order to track the deformation process. The system was calibrated for a 65x55mm field of view and resulted in spatial resolution equal to 27x27 $\mu m/pixel$. A Fluke Ti55 IR FlexCam operating in the 8 $\mu m$ to 14 $\mu m$ range which was set to have a thermal emissivity of 0.75 and an acquisition rate of 5fps was also utilized. The ambient temperature for IRT measurements was set to 21°C and the measurable surface temperature range was -25°C to 100°C with an accuracy of ±2°C.

## Results and Discussion

The objective of the tensile tests was to characterize the mechanical behavior and monitor the damage process in this type of FRPC specimens. The UTS and elongation of the specimen were measured to be 786 MPa and 2%, respectively. Note that the global strain in this figure was computed by averaging full field DIC data such as the ones shown in Fig.2b. Figure 2a shows the linear stress-strain behavior measured during these tests with an observed stress drop exactly at the UTS which was actually related to the final fracture of the sample visually verified. The correlated AE amplitude scatter plot (note that each dot in Fig.2a is computed by a corresponding waveform post-processed in real time by the AE data acquisition system) provides additional information about the damage accumulation process that cannot be inferred by the σ-ε curve. Specifically, it can be observed that the density and amplitude of recorded AE waveforms increased past the ~1% average global strain. In fact, at about 1.8% strain (point marked as 2 in the σ-ε curve), both the cumulative AE energy and average temperature change (Fig.2c) show abrupt increasing trends, which coincide with the drop seen in the σ-ε curve exactly at the UTS. It is worth mentioning that although AE and IRT provided in this test (as well as other similar tests not shown here) early damage warnings, which in the case of IRT further resulted in the identification of a localized "hot zone" (Fig.2d) later shown to correlate with the final failure area, the full field deformation monitoring approach did not indicate any significant longitudinal strain accumulations that could be used to track the onset and location of occurring damage, reconfirming that damage in composite materials could be barely visible up until final failure. At point 3 of the σ-ε curve, all three NDE methods provided average and full field indications of the occurred damage. In the case of DIC, damage information consisted of the identification of an area along the right edge of the specimen with different strains than the rest of the sample, which IRT also identified as relatively "colder". Post test visual inspection verified that this area corresponded to the failed zone of the tested specimen.

Post mortem one-to-one NDE correlations were attempted in order to obtain additional information about the capabilities of the NDE methods used to provide early damage pre-cursors and continuous damage monitoring for the tested FRPC tensile specimens. Specifically, Figure 3a presents three different NDE parameters (each corresponding to one of the three methods used) correlated as a function of time which in this case were represented by measured temporal changes in temperature measured by IRT. The mean local Poisson's ratio computed by calculating the average of Poisson ratio values defined locally by using at each available DIC measurement point longitudinal and transverse strains (reported in Fig2b and Fig.3b, respectively) showed a rapid rise at low temperature changes in the beginning of loading, and it remained nearly constant until point 3 (numbered points correspond to the ones in Fig.2a) where it rapidly increases, in agreement with the cumulative AE energy curve. The reason for this change can be found in the disproportions observed in measured transversal and longitudinal strains (again see Fig.2b and Fig.3b). This effect can be clearly shown in Fig.3b where local Poisson's ratio values are computed and demonstrate an evolution from a stochastic distribution at point 1 to a more percolated-like structure at point 3, which was later visually verified to coincide with locations where damage occurred. Furthermore, the calculated average shear strain evolution as a function of the temperature range was correlated to the AE cumulative energy. Fig.3c shows that the shear strain (y-axis values in microstrain) continuously increase and present a significant change in their slope near point 2, which as mentioned earlier was the point where the UTS was reached. The full field shear strain maps in Fig.3d provide supplementary evidence in the evolution of this DIC-defined parameter and shows that at point 3 significant shear strain accumulations have occurred.

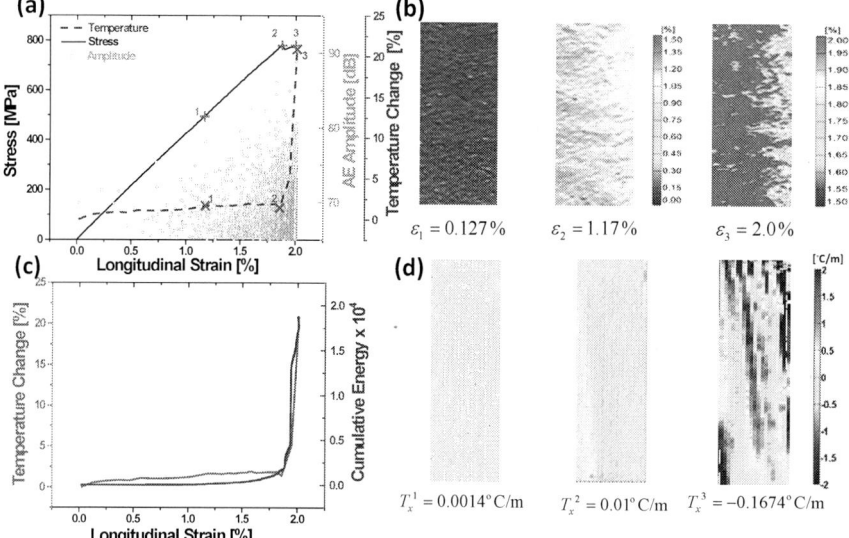

**Figure 2**: 
(a) Stress-average longitudinal strain (measured by DIC) curve Correlated with AE amplitude and average temperature change measured by IRT.
(b) Full-field longitudinal surface strains (the average of which resulted in the strain values reported in Fig.2a).
(c) Average temperature change evolution as a function of strain and correlated with cumulative AE energy.
(d) Full field transversal (x-axis) temperature gradient.

The assessment of damage related NDE information obtained by the tensile tests was evaluated by applying the same monitoring approach in fatigue tests (Fig.4a). Specifically, the implementation of the DIC method during fatigue testing enabled monitoring of the full field surface strain profile both during selected fatigue cycles for the entire specimen lifetime, as well as at predetermined points (e.g. peak load values) of each fatigue cycle throughout the experiment. Hysteresis stress (measured by the loading frame) strain (measured by averaging full field DIC data) loops typical in fatigue experiments were obtained (not shown here). Such loops can be used to define fatigue damage parameters such as the energy density (also referred as energy dissipation per cycle), residual stiffness (defined as the slopes connecting the two endpoints of each hysteresis loop), as well as extrema and mean strain values.

The correlation between the obtained DIC and AE data during fatigue tests assisted the identification of three distinct stages of damage evolution, as seen in Fig 4b. This damage evolution was further verified by evaluating the average Poisson's ratio values (computed as explained earlier by using full field strain data at the peak load) with respect to recorded AE amplitude distributions (Fig.4c) as a function of life fraction. The three stages of fatigue life are marked with dotted lines in Fig.4b-c. Stage I corresponded to initially high AE amplitude emissions, which however appeared to decrease as fatigue continued, and decreasing Poisson ratio values that can be explained by changes in the full field deformation patters reported in Fig.4a. In stage II, marked by the change in the slope of both the cumulative AE energy distribution in Fig.4b, as well as the corresponding change in the slope of the average Poisson ratio in Fig.4c, the recorded AE amplitudes are relatively low for the entire duration of this stage, while the Poisson's ratio increases slightly and then also remains nearly constant. The energy density during stage II presents a slight increasing trend which however dramatically changes at

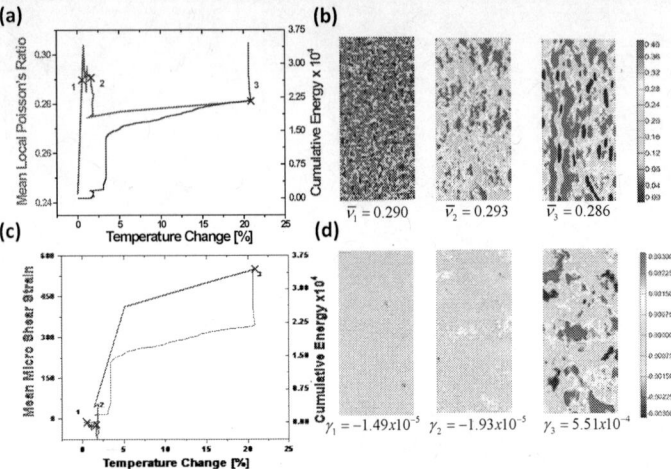

**Figure 3:** (a) Average Poisson's ratio and AE cumulative energy evolution as a function of temperature change.
(b) Local Poisson's ratio values at different stress level.
(c) Average shear strain and AE cumulative energy correlated with the temperature change, and
(d) full field shear strain values computed by DIC.

the onset of stage III further marked by a sudden increase of the average Poisson ratio values, followed by a significant increase of the AE amplitude and cumulative energy. Finally, the comparative exposition of full field surface maps for the change in temperature and transversal strain, as well as characteristic AE waveforms for each stage in the NDE results shown in Fig.5a demonstrates clear differences in the findings of each method for each fatigue stage, while the full field approaches further result in the early identification of the exact area where catastrophic failure occurred.

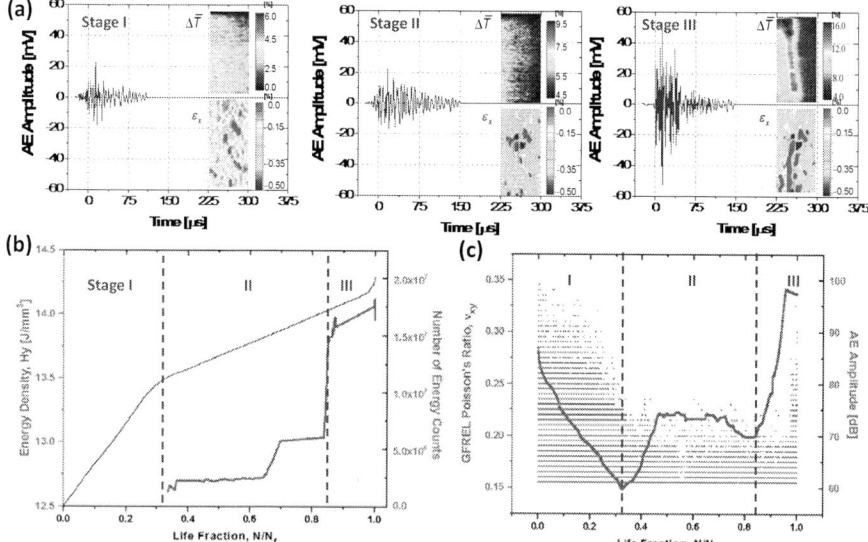

**Figure 4:** (a) Acoustic emission waveforms at each fatigue stage correlated with full field temperature change and transversal strains.
(b) Energy density and AE cumulative energy evolution, and
(c) average Poisson's ratio and AE amplitude distribution as a function of life fraction.

## Concluding Remarks

The data fusion nondestructive evaluation approach presented in this article shows great promise towards forming a physics-based framework towards monitoring, quantifying and modeling failure in composite materials. The results shown were based on both monotonic and fatigue tests and revealed the capability of the approach to not only flag incipient damage but to also correlate this information with actual mechanical behavior information in a cross-validated and damage specific way. Emphasis here was given on the overall damage states of the tested composites, since reliable identification of their evolution during fatigue could be the key factor in forming a damage-sensitive understanding of composite material mechanical and failure behavior.

## References

1. Saheb, N.D. and J.P. JOG, *Natural Fiber Polymer Composites : A Review.* Advances in Polymer Technology, 1999. **18**(4): p. 351-363.

2. Toldy, A., B. Szolnoki, and G. Marosi, *Flame retardancy of fibre-reinforced epoxy resin composites for aerospace applications.* Polymer Degradation and Stability, 2011. **96**(3): p. 371-376.

3. Kumar, S., et al., *Development and characterization of polymer–ceramic continuous fiber reinforced functionally graded composites for aerospace application.* Aerospace Science and Technology, (0).

4. Mandell, J.F., et al., *Analysis of SNL / MSU / DOE Fatigue Database Trends for Wind Turbine Blade Materials.* 2010(December).

5. Ghasemnejad, H., L. Occhineri, and D.T. Swift-Hook, *Post-buckling failure in multi-delaminated composite wind turbine blade materials.* Materials & Design, 2011. **32**(10): p. 5106-5112.

6. Hazeli, K., et al., *In situ identification of twin-related bands near yielding in a magnesium alloy.* Scripta Materialia, 2013. **68**(1): p. 83-86.

7. Phoenix, S.L., *Modeling the statistical lifetime of glass fiber/polymer matrix composites in tension.* Composite Structures, 2000. **48**: p. 19-29.

8. Talreja, R., *Damage and fatigue in composites – A personal account.* Composites Science and Technology, 2008. **68**(13): p. 2585-2591.

9. Tan, T. and C. Dharan, *Cyclic Hysteresis Evolution as a Damage Parameter for Notched Composite Laminates.* Journal of Composite Materials, 2010. **44**(16): p. 1977-1990.

10. Kaddour, A.S., et al., *16th International Conference on composite materials damage theories for fibre-reinforced polymer composites: the third World-Wide Failure Exercise (WWFE-III).* 2007.

11. Broughton, W.R., et al., *Assessment of Quasi-Static and Fatigue Loaded Notched GRP Laminates Using Digital Image Correlation.* Applied Mechanics and Materials, 2010. **24-25**: p. 407-412.

12. Giancane, S., et al., *Fatigue damage evolution of fiber reinforced composites with digital image correlation analysis.* Procedia Engineering, 2010. **2**(1): p. 1307-1315.

13. Bisagni, C. and C. Walters, *Experimental investigation of the damage propagation in composite specimens under biaxial loading.* Composite Structures, 2008. **85**(4): p. 293-310.

14. Marec, a., J.H. Thomas, and R. El Guerjouma, *Damage characterization of polymer-based composite materials: Multivariable analysis and wavelet transform for clustering acoustic emission data.* Mechanical Systems and Signal Processing, 2008. **22**(6): p. 1441-1464.

15. Loutas, T.H., et al., *Fatigue Damage Monitoring in Carbon Fiber Reinforced Polymers Using the Acousto-Ultrasonics Technique.* 2010.

16. Bentahar, M. and R. El Guerjouma, *Monitoring progressive damage in polymer-based composite using nonlinear dynamics and acoustic emission.* The Journal of the Acoustical Society of America, 2009. **125**(1): p. EL39-44.

17. Reis, P.F., Jose; Richardson Mel, *Fatigue Damage Characterization by NDT.* Applied Composite Materials, 2011. **18**: p. 409-419.

18. Bourchak, M., et al., *Acoustic emission energy as a fatigue damage parameter for CFRP composites.* International Journal of Fatigue, 2007. **29**(3): p. 457-470.

19. Innocenti, M., et al., *Properties of oxide/oxide CMCs for high temperature applications in gas turbines.* (I).

20. Fazzino, P.D., K.L. Reifsnider, and P. Majumdar, *Impedance spectroscopy for progressive damage analysis in woven composites.* Composites Science and Technology, 2009. **69**(11-12): p. 2008-2014.

21. Wang, D., et al., *Sensitivity of the two-dimensional electric potential/resistance method for damage monitoring in carbon fiber polymer-matrix composite.* Journal of Materials Science, 2006. **41**(15): p. 4839-4846.

22. Meo, M., U. Polimeno, and G. Zumpano, *Detecting Damage in Composite Material Using Nonlinear Elastic Wave Spectroscopy Methods.* Applied Composite Materials, 2008. **15**(3): p. 115-126.

23. Eliopoulos, E.N. and T.P. Philippidis, *A progressive damage simulation algorithm for GFRP composites under cyclic loading. Part II: FE implementation and model validation.* Composites Science and Technology, 2011. **71**(5): p. 750-757.

24. Van Paepegem, W., et al., *Monitoring quasi-static and cyclic fatigue damage in fibre-reinforced plastics by Poisson's ratio evolution.* International Journal of Fatigue, 2010. **32**(1): p. 184-196.

25. Kang, K.-W., D.-M. Lim, and J.-K. Kim, *Probabilistic analysis for the fatigue life of carbon/epoxy laminates.* Composite Structures, 2008. **85**(3): p. 258-264.

26. Allix, O. and L. Blanchard, *Mesomodeling of delamination: towards industrial applications.* Composites Science and Technology, 2006. **66**(6): p. 731-744.

# PROGRESSIVE FAILURE ANALYSIS OF POLYMER COMPOSITES USING A SYNERGISTIC DAMAGE MECHANICS METHODOLOGY

John Montesano[1], Chandra Veer Singh[1]

[1]Department of Materials Science and Engineering,
University of Toronto
184 College Street,
Suite 140, Toronto,
Canada, M5S 3E4

## ABSTRACT

A synergistic damage mechanics approach was implemented into a commercial finite element package to predict progressive failure of multidirectional polymer composite laminate components. The methodology retains the framework of continuum damage mechanics, while incorporating micromechanics to define the internal damage variables of the material system computationally. A user-defined subroutine was developed and damage evolution of quasi-static tensile loaded components was simulated using nonlinear analysis. The model predictions were in good agreement with experimental observations reported in the literature. Future considerations for model development should include incorporating the effects of multiaxial stresses on damage evolution, and adding additional damage modes to the model.

Keywords:  Polymer composites; damage mechanics; finite elements; multiscale modeling.

## 1. INTRODUCTION

One of the main benefits in using advanced fiber-reinforced polymer composites is that they can be tailored and optimized to suit a particular structural application by orienting the reinforcing fibers along multiple directions. A key issue with multidirectional composite laminates is their complex microstructure, which makes accurately predicting their mechanical behaviour difficult at best. This is specifically the case for progressive failure analysis of these materials since local cracks tend to develop in multiple directions simultaneously. In spite of the frequent use of multidirectional laminates in many industries including aerospace, automotive and wind energy, there is no rigorous and comprehensive analytical tool available for designers to assess their response in the presence of cracking in multiple plies. Consequently, components manufactured from multidirectional laminates are designed far too conservatively, which implies that the load-carrying capability of these materials are not fully utilized. The need to develop a computational tool for predicting durability and damage progression in multidirectional laminates is evident.

Current design approaches used in industry focus solely on strength failure, whereas in practical applications the failure characteristics may be defined by stiffness reduction or maximum component deflections. The reduction in the laminate stiffness is a result of progressive damage development. The damage mechanisms typically observed in multidirectional laminates begins with matrix cracking and fiber/matrix interface debonding, which are considered subcritical damage modes. Laminates can sustain a large number of such cracks before critical damage modes such as delamination and fiber fracture cause ultimate laminate failure. Understanding how the stiffness properties of a laminate structure vary with the changing applied load is therefore an important step in developing improved failure prediction models that are based on damage development of these material systems. Therefore, design procedures must account for ply level cracking and the resulting stiffness degradation behaviour.

Many studies have been reported in the literature that attempt to solve the problem of stiffness degradation due to ply cracking in multidirectional laminates. Some of the early work utilized one-dimensional shear lag models [1,2] while other studies considered two-dimensional variational approaches [3,4]. These models consider the variation of the local stress fields caused by the presence of matrix cracks. Reifsnider [5] developed the so-called critical element method, in which ply level cracking was considered to represent damage in both subcritical and critical elements. Another commonly adopted approach is to follow the so-called continuum damage mechanics (CDM) framework, which focuses on the evaluation of the overall stiffness properties of the damaged laminate through the homogenization of a representative volume element (RVE) [6-8]. This approach has been successful for a variety of laminate sequences, but is limited in application due to the need for the evaluation of damage constants through extensive experimental test programs. A synergistic damage mechanics (SDM) methodology has recently been utilized to address these issues in progressive failure analyses of multidirectional composite laminates containing multiple damage modes [9]. SDM retains the CDM framework at the mesoscopic level of the RVE, while incorporating microscopic cracking through tensor-valued internal damage variables that are ultimately determined through computational efforts in lieu of experimental tests. The SDM methodology has proven to be a powerful tool for the accurate prediction of progressive damage behaviour in multidirectional laminates [10].

In its present state, the SDM methodology has been utilized analytically to predict the progression of damage and the corresponding stiffness degradation for a variety of multidirectional laminates. However, the use of the SDM methodology for structural analysis has not yet been attempted. Therefore, the objective of this study is to implement the SDM prediction methodology into a commercial finite element (FE) package, and to predict the progressive failure of practical multidirectional laminate components. The aim is to develop a progressive failure analysis model based on the SDM approach so that it can be used as a viable tool by design engineers in industry. The subsequent section outlines the main theory related to the development of the SDM methodology, while the finite element implementation and the corresponding considerations will also be detailed. The results for uniaxially loaded laminate plates obtained using the FE model will be presented and compared to experimental data, while the prediction results for a plate with a central hole will also be presented. Finally, conclusions will be drawn regarding the modelling approach, and future work will be outlined.

## 2. SDM Model

Consider a multidirectional laminate consisting of both on-axis and off-axis plies as shown in Figure 1, where the off-axis ply orientations are denoted as $\theta_1$, $\theta_2$, etc. When such laminates are loaded in tension along the direction of the on-axis plies (i.e., axially), ply level matrix cracks will initiate in the off-axis plies first. Typically the 90° plies will develop transverse cracking the earliest due to their matrix-dominated ply orientation which results in brittle matrix fracture. On further application of the applied load, cracks will initiate and propagate in plies with other off-axis orientations resulting in multiple damage modes. The crack density in each of the off-axis plies continues to increase, and when the load reaches a certain critical value inter-laminar cracking (i.e., delamination) initiates. Delamination can eventually lead to the final failure of the composite structure, which comprises of fiber fractures and failure of the on-axis plies. The focus here is on ply level cracking and its effect on the laminate stiffness properties.

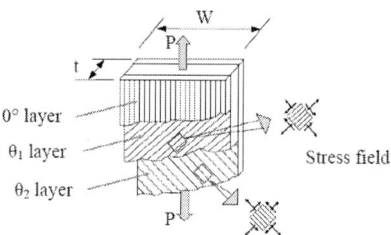

**Figure 1.** A multidirectional laminate loaded in axial tension.

### 2.1 Damage Stiffness Relations

The full details of the SDM methodology for predicting damage evolution in multidirectional laminates were reported by Singh and Talreja [9,10] thus only the key points of the analytical model will be presented here. The focus will be on damage analysis of
$[0_m / \pm\theta_n / 90_r]_s$ and $[0_m /90_r /\pm\theta_n]_s$ laminates with the intention that these laminates address many practical structures. Distinct ply-level matrix cracking scenarios will develop in the 90° plies as well as in the $\pm\theta$ plies for these specific laminates, thus there are a total of two damage modes considered for the model development.

By following the notations of CDM, the effects of evolving damage on the mechanical behaviour of a laminate can be analyzed by considering an RVE such as that shown in Figure 1. By considering the geometrical characteristics of the two corresponding damage modes, and through homogenization of the effects of these damage modes over the volume of the RVE, the overall laminate stiffness properties can be derived. The resulting damage field at a material point is described by a second-order damage tensor, $D$, that describes the damage state in an equivalent homogeneous continuum. Following the work of Talreja [11], the damage tensor components for a specific damage mode $\alpha$ are given by:

$$D_{ij}^{(\alpha)} = \frac{1}{V} \sum_{k_\alpha} \left( \int_S a n_i n_j dS \right)_{k_\alpha} \quad (1)$$

where $V$ is the volume of the RVE, $S$ is the surface area for a single crack, $k\alpha = 1, 2, \ldots, N$, and $N$ denotes the number of damage entities. The unit vector $n$ is normal to the crack surface and dictates the crack orientation, the components of which are defined by $n_i = (\sin\theta, \cos\theta, 0)$. The coefficient $a$ represents the extent of some influence normal to the crack faces, which can be defined by $\alpha = \kappa t_c$ [9]. The thickness of the cracked plies for a particular ply orientation is denoted by $t_C$, and $\kappa$ is the constraint parameter that accounts for the constraint on the corresponding matrix cracks caused by adjacent fibers or adjacent plies in the laminate. Thus with the assumption that all cracks have the same surface area, all cracks span the entire thickness of the corresponding ply, and the crack spacing $s^o{}_n$ presented in Equation 1 can be approximated by:

$$D_{ij}^{(\alpha)} = \frac{\kappa t_c^2}{s_n^o t} n_i n_j \quad (2)$$

As progressive damage takes place, the stiffness properties of the laminate are reduced. The corresponding laminate stiffness tensor, $C$, can be derived as [9]:

$$C_{pq} = C_{pq}^o + \sum_\alpha C_{pq}^{(\alpha)} \quad (3)$$

where $C^o{}_{pq}$ are the stiffness tensor components for the undamaged material, and $C^{(\alpha)}{}_{pq}$ are the changes in the stiffness tensor components caused by the corresponding damage mode, $\alpha$. For an orthotropic material with linear elastic stress-strain behaviour (i.e., $\sigma_p = C_{pq}\epsilon_q$) subjected to a plane stress state, the overall stiffness matrix is given by:

$$C_{pq} = \begin{bmatrix} \frac{E_x^o}{1-v_{xy}^o v_{yx}^o} & \frac{v_{xy}^o E_y^o}{1-v_{xy}^o v_{yx}^o} & 0 \\ \frac{v_{xy}^o E_y^o}{1-v_{xy}^o v_{yx}^o} & \frac{E_y^o}{1-v_{xy}^o v_{yx}^o} & 0 \\ 0 & 0 & G_{xy}^o \end{bmatrix} + \bar{D} \begin{bmatrix} 2a_1 & a_4 & 0 \\ a_4 & 2a_2 & 0 \\ 0 & 0 & 2a_3 \end{bmatrix} \quad (4)$$

The effective modulii for the undamaged material are denoted by terms $E_x^o$, $E_y^o$, $v_{xy}^o$, $G_{xy}^o$, and the $a_i$ are material constants to be determined from computational micromechanical FE models of the laminate. The effective damage parameter, $D$, can be defined for the specific laminate sequence. For a $[0m / 90r / \pm\theta n]$s laminate with cracking in the 90° and $\pm\theta$ plies, $D$ is defined by:

$$\bar{D} = \frac{2t_o^2}{t} \left[ \frac{1}{s_n^\theta} \frac{\kappa_\theta}{\kappa_{\theta=90}} \left( 2(2n+r)^2 \kappa_{90_{4n+2r}} - r^2 \kappa_{90} \right) + \frac{1}{s^{90}} \kappa_{90} r^2 \right] \quad (5)$$

where $t_O$ is the thickness of a single ply and $s^{90}$ is the crack spacing in the 90° plies. The constrain parameters that correspond to a particular damage mode are determined computationally using a micromechanical FE model, and are based on corresponding crack opening displacements [9].

### 2.2 Evolution of Ply Crack Density

The evolution of damage is predicted using a fracture mechanics based energy method which utilizes crack closure concepts. For example, the damage progression from State 1 with $N$ parallel off-axis cracks to State 2 where the cracks have multiplied to $2N$ will occur when the work required in going from State 1 to State 2 exceeds the critical energy release rate. Computational micromechanics is therefore used to evaluate crack surface displacement parameters as a function of crack densities (i.e., $1/s^\theta$ and $1/s^{90}$) using micromechanical FE models. The micromechanical model is therefore utilized to define the crack density-strain relationship (i.e., $\rho = \rho(\varepsilon)$). Further details are outlined in Ref. [10].

### 3. Finite Element Implementation

The analytical model described in Section 2 is implemented into the commercial FE software package ANSYS through its user programmable feature capabilities (i.e., a user-defined USERMAT subroutine). The user can define the material constitutive laws through the USERMAT Fortran-based algorithm, which can then be linked intrinsically to the ANSYS computational environment. For numerical computations, the USERMAT subroutine is called at every material integration point in the element mesh during every solution load step, and possibly multiple times during one load step for the Newton-Raphson convergence iterations. When called, the USERMAT subroutine requires the stresses, strains and the state variables (i.e., damage variables $D^{(\alpha)}$) at the beginning of the current load step for the corresponding material point, as well as the strain increments over the load step, which are determined by ANSYS as input. The subroutine is then required to provide as output the stresses at the end of the load step, which are determined using the degraded stiffness properties from the current strain state. The subroutine also determines the values of the state variables at the end of the current load step.

In this study, the form of the constitutive law is conducive for implementation into a displacement-based FE analysis scheme. Therefore, the developed USERMAT subroutine determines the stiffness matrix of the material using Equation 4, and subsequently determines the stresses at the end of a load step using the constitutive law, $\sigma_p = C_{pq}\epsilon_q$. The updated effective damage parameter, which is a function of the evolving crack densities, is determined using Equation 5. As indicated, the evolution of the crack densities with increasing strain is determined using micromechanics and is provided as input to the USERMAT subroutine, in the form of a curve, $\rho = \rho(\varepsilon)$, for each damage mode. Note that the constrain parameters, $\kappa$, and the material constants, $a_i$, are also provided as input for the USERMAT subroutine. As a result, the developed multiscale model is not a concurrent model, but is deemed a hierarchical model.

## 4. RESULTS AND DISCUSSION

The goal is to predict the evolution of damage in quasi-isotropic [0 / 90 / ±45]s glass fiber/epoxy laminates subjected to uniaxial quasi-static tensile loading using the ANSYS USERMAT subroutine discussed in Section 3. The results for a flat plate, which is analogous to a static test coupon, will first be presented to obtain the laminate stress-strain relationship (see Figure 2a for corresponding FE mesh). Then, the results for a plate with a central hole will be shown to highlight the capability of the model to predict damage development in a practical laminate structural component (see the one-quarter model shown in Figure 2b).

For the laminate studied, the ply properties used in the analysis are: $t_o$ = 0.5 mm, $E_1$ = 46 GPa, $E_2$ = 13 GPa, $G_{12}$ = 5 GPa, $v_{12}$ = 0.3. The material constants from Equation 4 are found to be $a_1$ = -1374.71 GPa, $a_2$ = -50.44 GPa, $a_3$ = 0, and $a_4$ = -720.96 GPa.

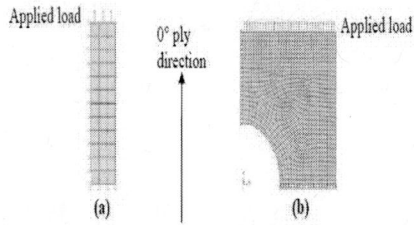

Figure 2.  FE mesh and boundary conditions for (a) flat coupon, (b) plate with central hole.

### 4.1 Stress-Strain Predictions

The flat plate FE model shown in Figure 2a utilizes PLANE 182 elements to model the homogenized laminate material system. Symmetry boundary conditions are used, and a nonlinear solution is performed to simulate a quasi-static uniaxial tensile load applied in the 0° ply direction as is shown. The predicted axial stress-strain curve is illustrated in Figure 3. The nonlinearity in the stress-strain behaviour caused by the combined effect of cracking in the 90° plies and the ±45° plies is effectively predicted by the SDM-based prediction model. The changes in the slope of the stress-strain curve correspond to the initiation of cracking in the 90° plies and the ±45° plies, as is expected [12]. The prediction model is capable with dealing this complex multimode damage scenario.

## 4.2 Component Analysis

The FE model of the plate with a central hole shown in Figure 2b also utilizes PLANE 182 elements. A nonlinear solution is also performed to simulate a quasi-static uniaxial tensile load applied in the 0° ply direction as is shown. Successive contour plots of the component crack densities at the indicated applied nominal stresses as predicted by the FE model are shown in Figures 4a and 4b. At an applied nominal stress of 68 MPa, cracking in the 90° plies initiates at the region of high stress concentration, whereas cracking in the ±45° plies has not yet initiated in the component. At a nominal stress of 118 MPa, the region of the component containing 90° ply cracks has increased in size and cracking in the ±45° plies has initiated as shown in Figure 4b. At an applied load of 200 MPa, cracking in the 90° plies is significantly more accelerated compared to the crack development in the ±45° plies, and the region of 90° crack development is much larger. The evolution of damage in the 90° and ±45° plies, and the resulting damage pattern predicted by the model, are consistent with the experimental findings of other notched quasi- isotropic laminates subjected to quasi-static tensile loading [13].

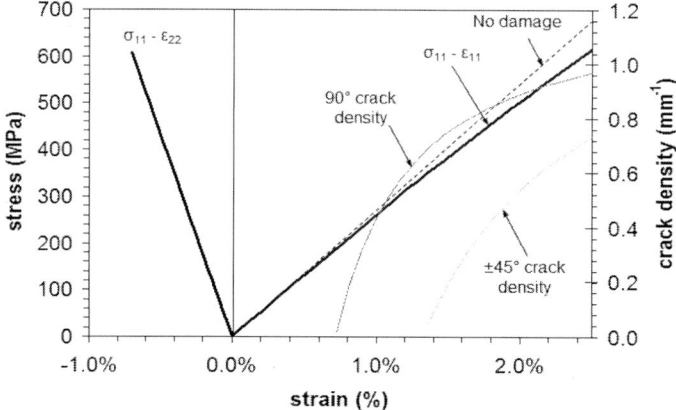

Figure 3.  Axial stress-strain curve, axial stress-transverse strain curve, 90° ply and ±45° ply crack density evolution.

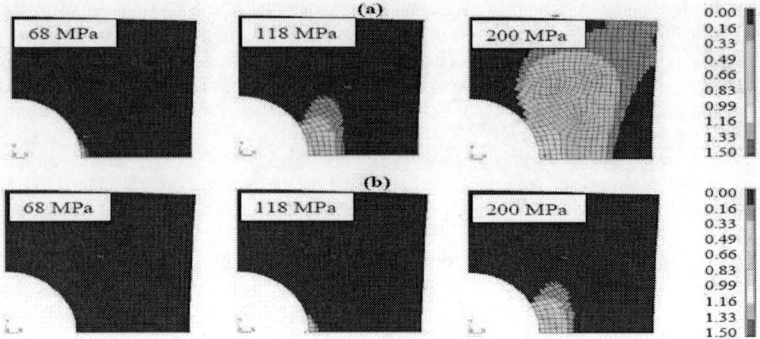

Figure 4. Successive contour plots of (a) 90° crack density (mm-1), (b) ±45° crack density (mm-1).

## 5. Conclusions

The synergistic damage mechanics approach was successfully implemented into a commercial finite element software package in order to predict the progressive failure of practical multidirectional laminate components, with the aim of developing a viable design tool for the industry. The results of the prediction model illustrate both the accuracy and practicality of the SDM methodology. The predicted intra-laminar crack progression for a notched quasi-isotropic laminate were in good agreement with experimental results for a similar laminate reported in the literature. Although the current state of the prediction model has yielded noteworthy results, future considerations for model development include incorporating the complex effects of multiaxial stresses on crack evolution. Additional damage modes such as 0° ply cracks, which are applicable for multiaxial stress states, and inter-ply delamination will also be considered in the prediction model. Finally, the addition of compressive damage modes, such as fiber kinking, will also be considered. The long-term goal is to employing the prediction model for simulating damage evolution of large-scale practical multidirectional laminate components subjected to cyclic loading, which may include wind turbine blades and aircraft structural components.

## References

[1] A.L. Highsmith, and K.L. Reifsnider. "Stiffness-Reduction Mechanisms in Composite Laminates," *Damage in Composite Materials*, ed. K.L. Reifsnider (Philadelphia, PA: ASTM STP 775, 1982), 103-117.

[2] S. Lim, and C. Hong, "Prediction of Transverse Cracking and Stiffness Reduction in Cross- ply Laminate Composites," *Journal of Composite Materials*, 23 (1989), 695-713.

[3] Z. Hashin. "Analysis of Cracked laminates: A Variational Approach," *Mechanics of Materials*, 4 (1985), 121-136.

[4] J.A. Nairn. "Strain Energy Release Rate of Composite Microcracking: A Variational Approach," *Journal of Composite Materials*, 23 (1989), 1106-1129.

[5] K.L. Reifsnider. "The Critical Element Model: A Modeling Philosophy," *Engineering Fracture Mechanics*, 25 (1986), 739-749.

[6] R. Talreja. "Transverse Cracking and Stiffness Reduction in Composite Laminates," *Journal of Composite Materials*, 19 (1985), 355-375.

[7] D.H. Allen, C.E. Harris, and S.E. Groves. "A Thermomechanical Constitutive Theory for Elastic Composites with Distributed Damage - I: Theoretical Development," *International Journal of Solids and Structures*, 23 (1987) 1301-1318.

[8] P. Ladeveze. "A Damage Computational Approach for Composites: Basic Aspects and Micromechanical Relations," *Computational Mechanics*, 17 (1995), 142-150.

[9] C.V. Singh, and R. Talreja. "A Synergistic Damage Mechanics Approach for Composite Laminates with Matrix Cracks in Multiple Orientations," *Mechanics of Materials*, 41 (2009), 954-968.

[10] C.V. Singh, and R. Talreja. "Evolution of Ply Cracks in Multidirectional Composite Laminates," *International Journal of Solids and Structures*, 47 (2010), 1338-1349.

[11] R. Talreja. "Continuum Modelling of Damage in Ceramic Matrix Composites," *Mechanics of Materials*, 12 (1991), 165-180.

[12] C.V. Singh, and R. Talreja. "A Synergistic Damage Mechanics Approach to Mechanical Response of Composite Laminates with Ply Cracks," *Journal of Composite Materials*, 47 (2013), 2475-2501.

[13] J. Xiao, and C. Bathias. "Damage and Fracture of Notched Non-Woven and Woven Composite Laminates," *Composites Science and Technology*, 52 (1994), 99-108.

# COMPUTATIONAL PREDICTION OF MECHANICAL PROPERTIES OF GLASSY POLYMER BLENDS AND THERMOSETS

David Rigby, Paul W. Saxe, Clive M. Freeman

Materials Design, Inc
343 West Manhattan Avenue
Santa Fe, NM 87501, USA

Benoit Leblanc

18 rue de Saisset, 92120 Montrouge, France

## Abstract

Atomistic simulations of the elastic constants of glassy polystyrene-poly(2,6-dimethyl-1,4- phenylene oxide) blends and 4,4'-diamino-diphenyl sulfone cured epoxy thermosets have been performed in order to examine the precision and accuracy currently achievable using molecular simulations. Bounds estimates obtained using moderately sized batches of independent amorphous structures have been shown to be comparable in magnitude to those obtained in most experiments. For the blend systems, the variation in tensile moduli with composition has been found to be very close to that observed experimentally, and rational explanations for the small
~16% discrepancy in absolute values have been given. In the case of the epoxy-based thermosets, good agreement with experimental moduli suggests that the approach used to create the crosslinked models leads to chemically and physically realistic models of these complex materials

**Keywords**: Atomistic Simulation, Thermosets, Polymer Blends, Mechanical Properties

## Introduction

In recent research, we have focused on use of atomistic simulations to calculate the elastic constants of glassy polymer blends and thermosets, which are two categories of polymer of interest in the aerospace, automotive, electronics and other industries. Mechanical property calculation of amorphous glassy polymers was first studied computationally in the pioneering work of Theodorou and Suter, who performed a detailed analysis of the entropic contributions to the elastic moduli, arriving at the important conclusion that the elastic moduli can be determined by considering only changes in potential energy on deformation [1]. The authors then proceeded to examine glassy atactic polypropylene, obtaining reasonably accurate predictions of the elastic constants within 15% of experimentally-measured values.

Following this early study, a number of workers applied the same methods to other glassy polymers, including polysulfone, bisphenol-A polycarbonate and polystyrene [2-5]. The accuracy of these later studies was lower overall than had been the case for polypropylene, and the precision reflected in the reported uncertainties was rather limited, with error bars in the range of ± 30-50%. Various hypotheses can be advanced to explain the lack of accuracy and precision, including the small system sizes used (~700-1200 atoms), limited sampling of the configuration space of the polymers through averaging only small numbers of independent structures, and reliance on force fields with fairly low accuracy, as supplied with the commercial software packages used for the computations. Most importantly however, the question of how to analyze the elastic constants measured *in silico* using collections of nanoscopic domains, to obtain estimates of the precision – or bounds – of the moduli of macroscopic material had not at this time been considered for polymeric systems.

This situation was addressed by Suter and Eichinger, who pointed out that estimates of the extreme upper and lower bounds as obtained by averaging stiffness and compliance matrices (the so-called Voigt and Reuss bounds) are too wide to be useful for comparing materials when obtained using such small disordered systems [6]. These authors reasoned that the nanoscopic domains, or 'cells', used in atomistic models of amorphous polymers can be viewed as heterogeneous domains embedded within macroscopically homogeneous material, and proceeded to apply the methods of Hill and Walpole developed earlier for treating composite material mechanical properties [7, 8]. In this way they were able to demonstrate significantly improved estimates of the bounds on elastic constants of glassy polymers obtained using ensembles of atomistic-scale models whose size is limited to a few nanometers. This development in principle makes it possible to quantify small differences between materials associated with chemical structure, or composition in the case of blends.

In recent work, we have performed calculations on a variety of amorphous glassy polymers using the Theodorou-Suter mechanical property approach augmented with the Hill-Walpole analysis method. Since there is a lack of published systematic studies of polymer moduli determined by atomistic methods, we have specifically selected an engineering polymer material consisting of a compatible blend of two polymers –

polystyrene and poly(2,6-dimethyl-1,4-phenylene oxide) – for which tensile moduli have been measured as a function of composition, and for which the pure component tensile moduli differ only marginally, by less than 20%. Such a system provides an almost ideal model for testing both the accuracy and precision of computational predictions. Following this study we have proceeded to apply the method to a second category of polymeric materials in the form of densely crosslinked glassy epoxy resins used as matrices in modern composites. Since it is known experimentally that moduli can be sensitive to the molecular topology of the base resin, three systems differing in the architecture and chemical functionality of the resin component were selected for the investigation. The following sections describe the modeling procedures used and compare predicted results for the two types of polymer system with experimentally-measured values.

## Model Building and Mechanical Property Simulation Procedures

The model building and simulation tools employed are those provided within the MedeA® simulation environment [9]. Specifically, these tools provide for creation of realistic model configurations of amorphous polymers, polymer blends and the low molecular weight mixtures of components present in epoxy thermosets prior to curing.

For molecular mechanics and dynamics simulation, the MedeA® environment also makes use of the powerful LAMMPS program developed by Plimpton and coworkers at Sandia National Laboratories [10]. Thus, input models and command scripts are prepared and submitted to LAMMPS, with raw data returned to MedeA for further processing including, in the present work, application of the Hill-Walpole method to obtain bounds on elastic constants based on analysis of batches of independent configurations of the amorphous glasses. All calculations have used the PCFF+ all-atom force field, which is a based on the CFF/PCFF class II force fields included with the LAMMPS distribution, following extensive refinement of nonbond parameters using liquid state PVT and cohesive property data for small molecules [11, 12]. Batch sizes of ~100 independent configurations were used throughout most of the work.

For thermoset building, we have used an algorithm designed to generate relatively strain-free crosslinked models with a high degree of crosslinking. The algorithm combines approaches used in two separate areas of polymer modeling, the first of which was introduced by Eichinger and coworkers in the late 1980's for Monte Carlo based prediction of the *topology* of lightly crosslinked elastomer systems [13]. These authors introduced the concept of a capture sphere surrounding all possible crosslinking points. Pairs of sites falling within the capture sphere were then considered as candidates for formation of crosslinks, leading to branch formation and eventual formation of the infinite network as the capture sphere radius was slowly incremented [14]. In our thermoset work, which differs from the elastomer modeling in its creation of atomistic ally-detailed structures and high density of crosslinking associated with thermosets, we begin with a completely unreacted mixture and a relatively small capture sphere radius of around 0.4 nm and form bonds between all pairs of reactable groups lying within the capture spheres of each site (limited, as appropriate, by the maximum number of reactions each site is capable of undergoing). Since 'all-atom' models are being constructed, and since the initial bond formation step creates unrealistically long bonds and distorts other features of the local molecular geometry such as valence angles, we follow the bond formation with a second step, widely used to relax models of amorphous polymers since the early 1990's, and consisting straightforwardly of a small number of iterations of energy minimization and molecular dynamics [15]. Each iteration includes approximately 1000 steps of molecular dynamics, with the full relaxation after a crosslinking cycle requiring a few tens of CPU seconds for a model with, say, 2500-5000 atoms.

Following the relaxation, the entire structure is subjected to testing to ensure that the growing network model remains chemically realistic, which is achieved by computing metrics such as the maximum strain of any bond in the system, and by testing to ensure that neither the bond formation itself nor the subsequent relaxation has introduced defects such as ring catenation, in which pairs of aromatic or aliphatic rings become unrealistically interlinked. In the relatively rare event that such an unphysical structure has formed, the crosslinking is reversed and high temperature molecular dynamics is applied to the system before repeating the crosslinking and relaxation process.

After each crosslinking cycle, the neighbor lists associated with each capture sphere are regenerated prior to commencement of a new cycle. When a point is reached at which all possible reactions at a given sphere radius have been completed, the capture sphere radius is incremented, usually by .05 nm, and the process is continued until either a desired total extent of reaction has been reached or until a defect in the structure limits further reactions. Experience using the thermoset building algorithm has shown that high degree of conversion is possible without introduction of significant bond strain. Typically, the maximum acceptable bond strain is set at a moderate level of 0.1 (i.e. 10%), though frequently the observed maximum strain remains quite low – often less than 5% - to quite high extents of reaction, as illustrated in figure 1 below.

For elastic constant calculation, we have again made use of the LAMMPS program. In order to obtain the symmetric 6x6 array of elastic constants (resulting from use of the Voigt 6x1 stress and strain vector notation) a total of 13 LAMMPS energy minimizations are performed, with the first using the original equilibrated undeformed structure, and the remaining twelve performed after applying six tensile and six shear deformations. Strains of ±0.001 are used, with the reminimization energy convergence criterion set to $1.0 \times 10^{-6}$ kcal/mol/Å. The resulting $\Delta\sigma_i/\Delta\varepsilon_j$ values then yield the individual elastic constants and associated tensile, shear and bulk moduli and Poisson's ratio.

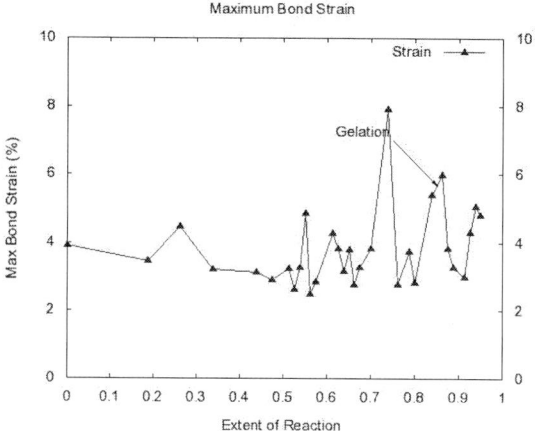

**Figure 1.** Maximum bond strain as a function of extent of reaction for a typical epoxy crosslinking process.

## Systems Studied

### Blends of Poly(styrene) and Poly(2,6-dimethyl 1,4-phenylene oxide)

Poly(styrene)-Poly(2,6-dimethyl-1,4-phenylene oxide) (PS-PPO) is a well-known blend system in which the two components are miscible in all proportions. The tensile moduli of PS-PPO blends have been studied experimentally by Kleiner and coworkers, using several polystyrenes with molecular weights ranging from ~4000 to $2 \times 10^6$, and three PPO samples with molecular weights of 17000, 35000 and 54000 [16]. Measurements using injection molded tensile specimens were performed for pure polymer and three blend systems containing approximately 25, 50 and 75 weight percent of PPO.

Since the high molecular weight systems for which experimental results were given by Kleiner et al. are difficult or impossible to model directly (e.g., a model containing just a single molecule of the highest molecular weight polystyrene would contain ~$3.2 \times 10^5$ atoms) the simulations have used shorter chains – with most model systems containing five polymer chains each with a degree of polymerization equal to 30, giving a total system size of ~2500 atoms. As was the case with the experimental studies, the compositions ranged from pure PS to pure PPO, with PS:PPO ratios of (5:0), (4:1), (3:2), (2:3), (1:4) and (0:5) respectively. In addition, this initial research has utilized densities for the pure PS systems of 1.033 gcm$^{-3}$, as appropriate for the ambient temperature density of an actual ~3000 molecular weight polymer, which is about 1% below the high molecular weight polystyrene value of slightly above 1.04 gcm$^{-3}$. Experimental data for the molecular weight dependence of PPO, and the value appropriate for the ~3600 molecular weight sample studied do not appear to be available and consequently the value of 1.06 gcm$^{-3}$ quoted for commercial samples of higher molecular weight PPO was used. Finally, for the blend systems, densities were calculated assuming additivity of the molecular volumes of each component.

**Epoxy Thermosets**

As noted above, epoxy thermosets, often formed by reacting a tetrafunctional amine with an epoxide, are widely used in advanced composite applications where they are combined with fiber or particulate reinforcing material. The present study has focused on one particular aspect of epoxy-based matrix material, namely the effect of varying the chemical functionality of the resin component on the small strain elastic constants of the crosslinked resin. Specifically, using the standard nomenclature $RA_{fa}+R'B_{fb}$ polymerization types applied to such reactions, where fa and fb denote the functionality of individual reactants, we concentrate on comparing the elastic constants of $RA_4+R'B_2$, $RA_4+R'B_3$ and $RA_4+R'B_4$ type systems, which have been studied experimentally by White et al. and by Behzadi and Jones [17, 18]

The cross linker (or curing agent, or hardener) used in this work is diaminodiphenylsulfone (DDS) depicted in the form of the 4,4' isomer in figure 2.

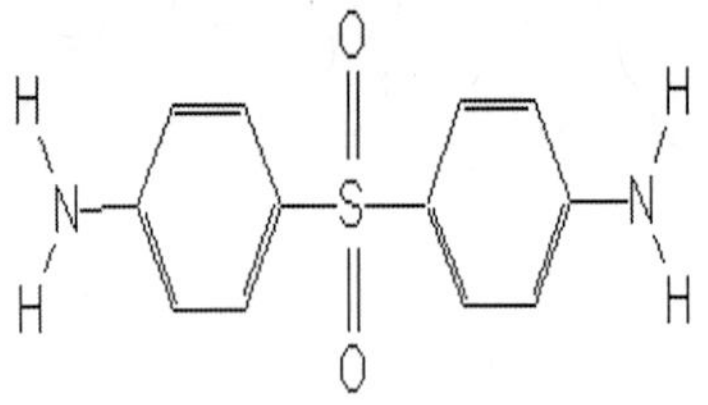

Figure 2.    4,4'-DDS curing agent

The resins used, diglycidyl ether of bisphenol A (DGEBA), triglycidyl p-aminophenol (TGAP) and tetraglycidyl diaminodiphenylmethane (TGDDM) are chemically similar to each other but contain different numbers of glycidyl ether end groups capable of reacting with the amine, as shown in Figure 3 below.

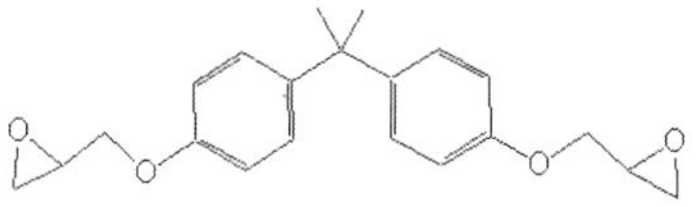

a. DGEBA

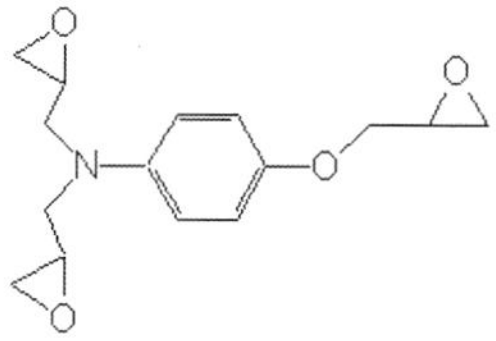

b. TGAP

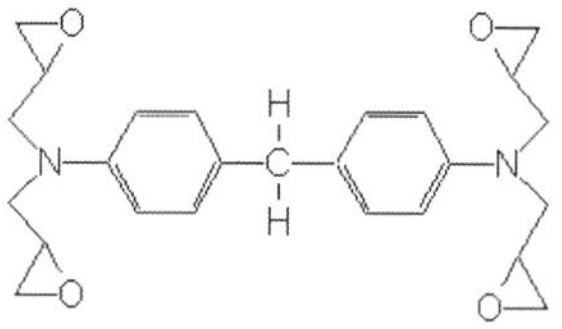

c. TGDDM

**Figure 3.** Di, tri and tetrafunctional resins, (a) DGEBA, (b) TGAP and (c) TGDDM.

For modeling the crosslinking reaction, it can be noted that the chemical reaction between a primary amine and the reactive oxirane rings in the resin is as follows:

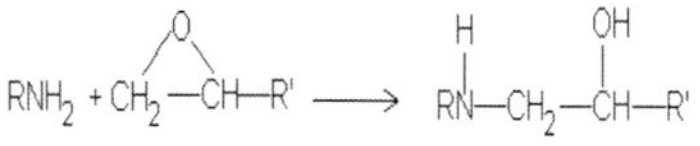

**Figure 4.** Reaction between the primary amine and oxirane ring.

A further reaction with the secondary amine group can then occur, giving a fully-reacted product with a tertiary amine structure, i.e.

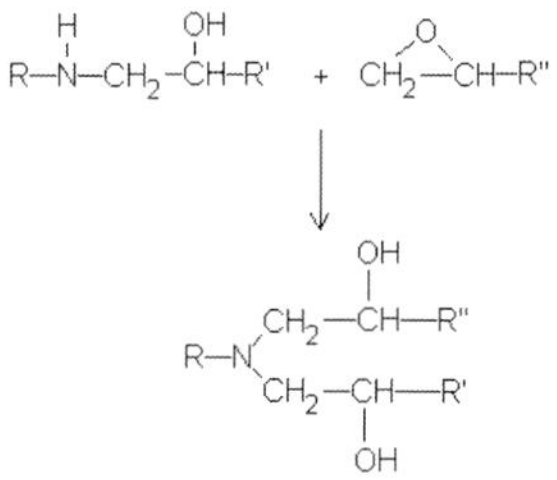

**Figure 5.** Second crosslinking reaction involving secondary amine groups.

This implies that in order to obtain 100% reaction of all functional groups when preparing epoxy resins based on DGEBA and a diamine, a ratio of two epoxy to one amine moieties is required.

For molecular modeling and simulation purposes it is convenient to work with 'network segments', in a manner similar to that used when building models of ordinary polymers where repeat units as opposed to actual monomers are generally used. In the case of the present epoxy- amine reactions, the oxirane rings of the resin are opened as shown in figure 6, creating the terminal methyl shown in bold type, such that the crosslinking reaction can then be completed by linking the terminal carbon with an amine nitrogen, followed by deletion of hydrogen atoms.

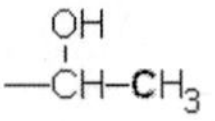

$$-CH(OH)-CH_3$$

**Figure 6.** *Network segment* epoxy units used in the uncured cross-linkable system.

## Results and Discussion

**Preliminary Validation Studies**

Before embarking on the extensive model building, simulation and mechanical property analysis required by the PS-PPO blends study, a series of calculations was performed to assess the accuracy of the force field by computing the densities of small molecules appearing as fragments within the polymers. Specifically, densities were calculated over an approximately 200 degree range of temperatures beginning at 300K for ethylbenzene and diphenyl ether. Calculations at each temperature consisted of constant pressure (NPT) simulations of duration 200ps, using a pressure of 1.01325 kPa (1 atm). The calculated densities, illustrated in figures 7 and 8, show overall excellent agreement with experimental data obtained from the DIPPR database with a typical error of ~0.7% for ethylbenzene at 323K [19].

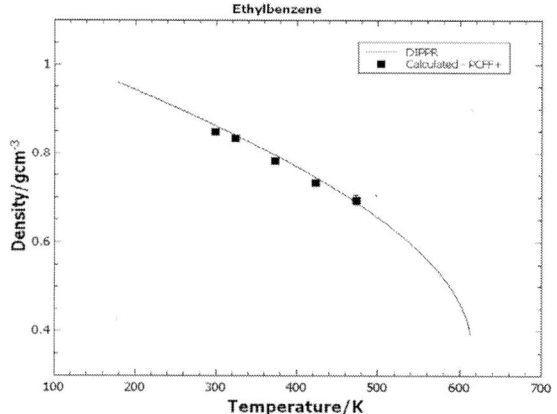

Figure 7.　　Ethylbenzene densities calculated using PCFF+

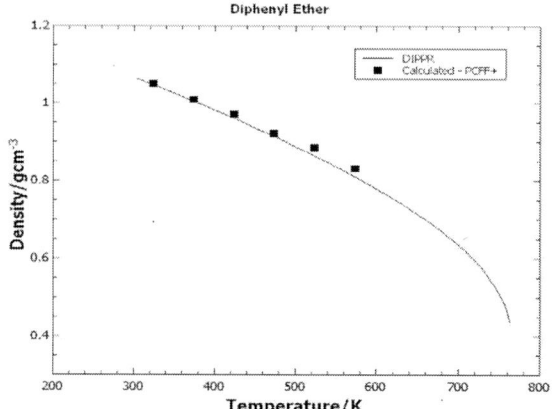

Figure 8.　　Diphenyl ether densities calculated using PCFF+.

## PS-PPO Moduli

As discussed above, the distributions of raw elastic constant data obtained for amorphous models with dimensions of around 0.3 to 0.4 nm are invariably rather broad, as illustrated in figure 9 which depicts the tensile moduli obtained for approximately 100 independently-generated models of the pure PPO system. After application of the Hill-Walpole analysis method however, the lower and upper bounds on the tensile modulus of the macroscopic bulk material are estimated as 2.226 GPa and 2.272 GPa respectively. Similarly narrow bounds are also obtained for the pure PS and for the blend systems, as shown in figure 10, which compares the calculated tensile moduli with the experimental data of Kleiner et al. measured on higher molecular weight polymers.

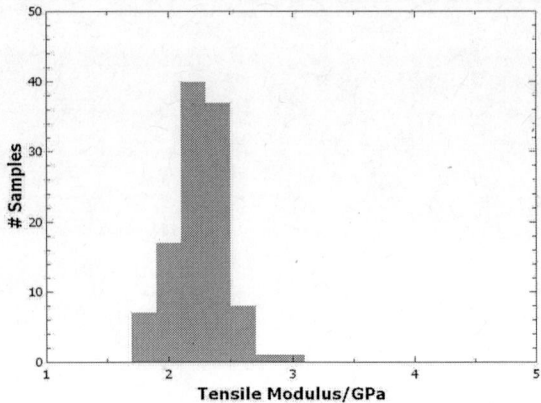

**Figure 9.** Distribution of tensile moduli calculated for a batch of 100 independently-generated PPO configurations.

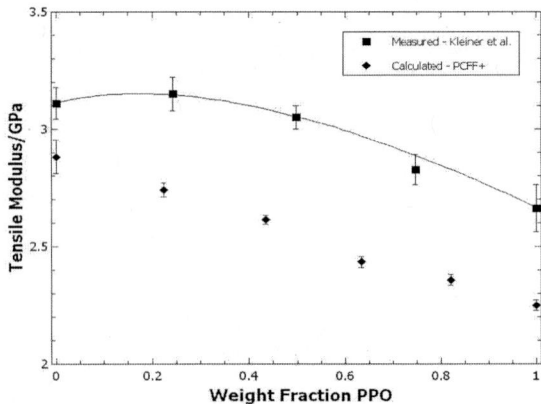

**Figure 10.** Calculated and experimentally-measured tensile moduli of PS-PPO blends. Squares - using experimental data from fig. 1 of ref. 16. Diamonds – calculated.

Two conclusions are immediately apparent on inspection of figure 10. Firstly, the bounds estimates obtained from the atomistic modeling as indicated by the error bars are comparable with, and actually slightly smaller than, the experimental uncertainties. Secondly, the percentage decrease in the modulus as the composition of the blend changes from pure PS to PPO is accurately reflected by the modeling.

Additional features not yet accurately reproduced in the modeling results are that the absolute values of the calculated moduli are lower than the experimental values (e.g., by about 16% for the PPO), and that there are no indications of the apparent synergistic effects that result in higher moduli for the blends than would be expected based on simple linear composition dependence. It is probable that both of these discrepancies are associated with dependence of the modulus on density. In the first case, as noted earlier, the present calculations assumed a lower density for the polystyrene than is known to apply to high molecular weight material, and also used a possibly less than accurate canonical handbook value for PPO. Moreover, use of relatively short chains also introduces a relatively high concentration of chain ends into the models which does not exist in the experimental systems, a factor which should also decrease the moduli of the model systems. In the case of synergistic effects, it can be recalled that the density used in simulations of the blend systems was assumed to be based on additivity of molecular volumes, whereas in the actual experimental systems the density is higher than expected using this assumption. Computationally, estimating the density of glassy polymers with high precision is a challenge due to exceptionally slow relaxation times. However, using experimentally measured high polymer densities and larger atomistic models containing longer polymer chains remains within reach using modern parallel computing facilities, and would provide an interesting future study, as would a more basic study of sensitivity of moduli to density, which will be addressed in future work in this area.

### Thermoset Moduli

Since the crosslinking process is essentially random, albeit influenced by correlations between groups arising from intermolecular packing requirements, a distribution of extents of reaction and network topologies results from the building process. A typical distribution of extents of reaction is illustrated in figure 11 for a DDS-cured system, where it is observed that the degree of crosslinking in the majority of systems studied is in the range ~80-95%.

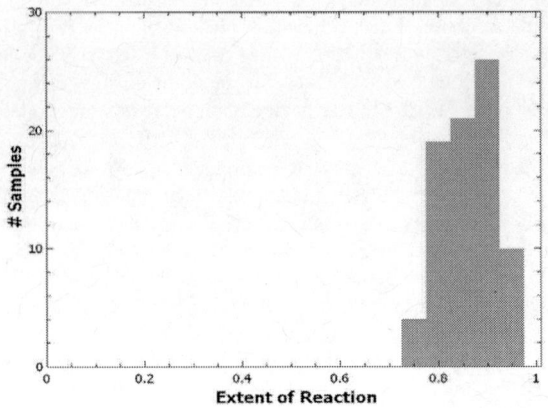

**Figure 11.** Typical distribution of degree of crosslinking in DGEBA resin models crosslinked using 4,4' DDS

The associated distribution of thermoset tensile moduli based on raw data from a large number of models is shown in figure 12. Here it is observed that the distribution is comparable with but perhaps slightly broader than the corresponding distribution obtained using PPO shown in figure 9, which is expected owing to the fact that the thermoset samples vary both in bonding topology and in the number of crosslinks.

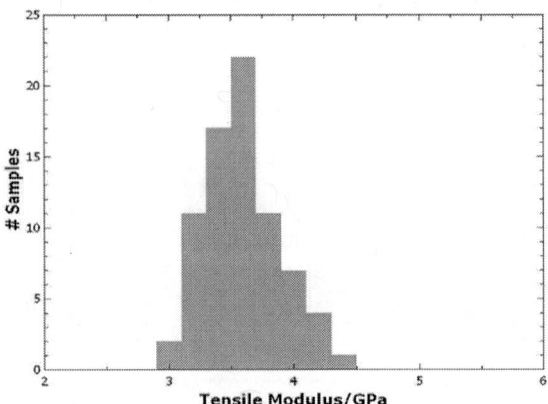

**Figure 12.** Distibution of tensile moduli of a batch of cross-linked DGEBA-DDS thermoset models.

Results obtained from application of the Hill-Walpole analysis to the thermoset systems based on the three different resin architectures are listed in table I, where the computed values are compared with the experimentally-reported data. Note that the Theodorou-Suter method used for the modulus calculations effectively corresponds to a high frequency measurement and consequently the experimental data quoted for TGAP and TGDDM in the table corresponds to the experimental results at the highest of three frequencies examined in ref. 18.

Table I. Comparison of calculated and experimentally-measured tensile moduli for epoxy resins based on di, tri and tetrafunctional resin architectures.

| Resin | Calculated (GPa) | Experiment (GPa) | Ref. |
|---|---|---|---|
| DGEBA | 3.49-3.53 | 2.4±.15-3.2±.15 | 17 |
| TGAP | 4.42-4.45 | 4.396±.027 | 18 |
| TGDDM | 5.18-5.19 | 5.103±.033 | 18 |

As with the PS, PPO and PS-PPO blend systems the modulus bounds estimates are narrow and well-defined, comparable with experiment for the TGAP and TGDDM systems and considerably narrower than those estimated from the dynamic mechanical analysis measurements of ref. 17. Quantitatively, agreement with experiment is excellent for the TGAP and TGDDM systems, but less so for the materials formed from DGEBA. In the latter case the experimental range quoted in table 1 applies to 1 Hz storage modulus measurements performed as a function of estimated degree of cure, which suggest that the most appropriate value for comparison purposes, corresponding to 80-90% cure, would be around 2.6±.15 GPa. Possible explanations for the lower experimental values compared with the simulations include the fact that the commercial DGEBA may contain molecules with more than the single bisphenol A moiety assumed in the simulation work which would be expected to lower both the density and the number of covalent crosslinks per unit volume, both of which may cause lowering of the modulus. As was the case in the PS-PPO study a future study of density effects is indicated, as also is a study of the exact chemical structure of the DGEBA.

## Summary and Concluding Remarks

The present investigation of the accuracy and precision of elastic constant measurements of glassy amorphous polymers using atomistic level simulations clearly demonstrates that improvements in simulation methodology and computer hardware developed over the past decade are in the process transforming this area of materials modeling into a routine activity. Detailed analysis of raw data from simulations on moderate sized batches of configurations of small model systems containing just a few thousand atoms has been shown using both the PS- PPO engineering polymer blend system and epoxy thermosets to yield a precision comparable with experiment. The reliability of the estimates also makes it possible to probe the origins of remaining relatively small discrepancies between experimental and simulation data. Finally, the successful study of the properties of thermosets provides a good indication that the method used to construct the thermoset models is capable of yielding chemically and physically realistic structures suitable for use in studying other important properties of such materials such as adhesive and cohesive behavior in the presence of solid substrates such as carbon fibers or hydroxylated metal surfaces, which will be addressed in future studies.

### Acknowledgement

The authors express their gratitude to Professor B.E. Eichinger of the University of Washington for providing guidance on implementation of the Hill-Walpole averaging method.

## References

1. D.N. Theodorou and U.W. Suter, Macromolecules, 19 (1986), 139-154.

2. C.F. Fan and S.L. Hsu, Macromolecules, 25 (1992), 266-270.

3. M. Hutnik, A.S. Argon, and U.W. Suter, Macromolecules, 26 (1993), 1097-1108.

4. C.F. Fan, T. Cagin, Z.M. Chen, and K.A. Smith, Macromolecules, 27 (1994), 2383-2391.

5. T. Raaska, S. Niemela, and F. Sundholm, Macromolecules, 27 (1994), 5751-5757.

6. Suter, U.W. and Eichinger, B.E., Polymer, 43 (2002), 575-582.

7. R. Hill, J. Mech. Phys. Solids, 13 (1965), 213-222.

8. L.J. Walpole, J. Mech. Phys. Solids, 14 (1966), 151-162.

9. MedeA® Simulation Environment, Materials Design, Inc., Angel Fire, NM, USA.

10. S.J. Plimpton, J. Comp. Phys., 117 (1995), 1.

11. H. Sun, S.J. Mumby, J.R. Maple and A.T. Hagler, J. Am. Chem. Soc., 116 (1994), 2978- 2987.

12. D. Rigby, Materials Design, Inc. unpublished.

13. See, for example, L.Y. Shy, Y.K. Leung, and B.E. Eichinger, Macromolecules, 18 (1985), 983-986.

14. O. Akgiray, Die Makromolekulare Chemie: Macromolecular Symposia, 76 (1993), 211- 218.

15. See *Polymer User Guide – Amorphous_Cell*, Chapter 3 (Biosym Technologies, San Diego, CA, 1991-1995).

16. L.W. Kleiner, F.E. Karasz, and W.J. MacKnight, Polymer Eng. Sci., 19 (1979), 519-524.

17. S.R. White, P.T. Mather, and M.J. Smith, Polym. Eng. Sci., 42 (2002), 51-67.

18. S. Behzadi and F.R. Jones, J. Macromol. Sci., Part B: Physics, 44 (2005), 993-1005.

19. DIPPR 801 Thermophysical Property Database, http://www.aiche.org/dippr.

Advanced Composites for Aerospace, Marine, and Land Applications
Edited by: Tomoko Sano, T.S. Srivatsan, and Michael W. Peretti
TMS (The Minerals, Metals & Materials Society), 2014

# MULTI SCALE CHARACTERIZATION OF SiC/SiC COMPOSITE MATERIALS

D. Frazer[1], M.D. Abad[1], C. Back[2], C. Deck[2], P. Hosemann[1]

[1] University of California Berkeley
Etcheverry Hall
94720 Berkeley, CA, USA

[2] General Atomics
San Diego, CA, USA

**Abstract**

SiC fiber-reinforced SiC matrix composites (SiC/SiC) are under consideration as a structural material for a range of nuclear applications. While these materials have been studied for decades, recently new small scale materials testing techniques have emerged which can be used to characterize SiC/SiC materials from a new perspective. In this work cross section nanoindentation was performed on SiC/SiC composites revealing that both the hardness and Young's modulus was substantially lower in the fiber compared to the matrix despite both being SiC. Using a Scanning Electron Microscopy (SEM) it was observed that the grain growth of the matrix during formation was radially out from the fiber with a changing grain structure as a function of radius from the fiber center. Focused ion beam machining was used to manufacture micro-cantilever samples and evaluate the fracture toughness and fracture strength in the matrix as a function of grain orientation in the matrix.

Keywords:   SiC, SiC fibers reinforced SiC matrix, nanoindentation, FIB, Micro cantilevers testing

## Introduction

Fission and fusion reactors are attractive options to meet the energy demands of our economy and society by providing economically competitive stable core energy generation while having a minimal detrimental impact on our environment [1,2]. Due to these attributes there have been many developmental efforts worldwide to improve and design new types of reactors [1,2]. Many advanced reactor concepts are considering significantly higher operating temperatures for process heat production and also more efficient thermodynamic cycles [3-5]. Both call for more suitable materials for high temperature applications. In addition the recent Fukushima events have illustrated that materials with no exothermic reaction or hydrogen production in steam in the event of a Loss Of Coolant Accident (LOCA) need be evaluated. Due to the excellent thermal conductivity, the low reactivity with water and steam as well as the high strength at elevated temperatures, and low neutron cross section SiC is being considered as a structural material for a range of these advanced reactor concepts [6-9]. The excellent thermal conductivity of SiC/SiC would allow easy heat transfer from the nuclear fuel to the coolant of the reactor providing higher energy conversion efficiency while maintaining excellent structural and neutronic performance [10]. Pure SiC does have the challenges of low ductility and no tolerance against brittle fracture therefore SiC matrix SiC fiber composites are being considered to mitigate these challenges [11, 12]. The use of SiC/SiC composites in reactor applications would provide superb safety margins for high temperature stability and radiation resistance to neutrons [13]. The SiC/SiC composites materials have excellent chemical compatibility with the gas (He) and liquid (Li,Pb) coolants proposed for fusion and advanced fission reactors [14-16].

However, SiC/SiC composites have challenges. While better than pure SiC in terms of fracture behavior [11,12,17] it does not have the ductility and fracture toughness of a metal and design challenges related to joining and material handling remain in applying this material to meet nuclear reactor needs within a radiation environment [18]. One of the most critical challenges is achieving consistent mechanical properties over the 14 ft long fuel rod. As for all composites the mechanical properties of a composite structure is determined by the interplay of the individual components; in this particular case by the interplay of fibers with the surrounding matrix material [8,19]. This is facilitated by an interface layer, such as pyrolytic carbon, located between the fibers and matrix. Reinforcing fibers coated with an interface layer, can bridge cracks, inhibit crack propagation, and slide or pull out of the matrix, affording improved mechanical performance [11,13,14]. However, as interfacial shear strength improves the overall composite strength, it can reduce the toughness due to decreased fiber sliding and pullout [20,21]. Various composite designs attempt to balance these toughening and strengthening characteristics in order to optimize material properties for nuclear applications. While to engineers obviously the strength and performance of the entire SiC/SiC composite structure is of interest, it can help structural modeling efforts to know more of the behavior of individual components. In addition it can be interesting to compare the properties of the deposited SiC to the properties of bulk SiC as well as investigate the properties of the fibers themselves. The intrinsic features of SiC make SiC fiber-reinforced SiC matrix composites attractive structural materials for nuclear applications, since SiC/SiC composites can overcome the inherently brittle facture behavior of monolithic SiC ceramics [20].

In this research we study the separate mechanical properties derived from the fiber and the matrix of the SiC/SiC composites. It is the purpose of this work to demonstrate that modern nanoscale mechanical testing techniques like nanoindentation and micro cantilever testing in combination with microstructural characterization techniques (Raman spectroscopy and SEM) allow insight about the properties of individual composite constituents leading to an understanding of the property of each component individually and acquire a better understanding about their interplay as a composite and identifying the weakest link. Small scale mechanical testing is required in order to test the matrix and fibers separately in the as manufactured composite. This work is a first step toward understanding the materials intrinsic properties rather than the extrinsic property as a compound.

**Experimental part**

Two different developmental composite samples, GA-1 and GA-2, were provided to the University of California Berkeley (UCB) by General Atomics (GA). These samples are not fully infiltrated, however, they are enable assessment of these methods, both specimens utilize commercially available nuclear grade stoichiometric SiC fibers (Tyranno SA3) which have excellent thermal conductivity [22]. The matrix of the SiC/SiC composite was processed using the chemical vapor infiltration (CVI) method which has the advantage of low thermal and mechanical stress during the densification process [19,23], and produces a stoichiometric, crystalline β-SiC which is highly stable to neutron irradiation [8]. The SiC/SiC composites were polished using standard silicon carbide papers up to 4000 grit followed by polish with 1 μm diamond grit.

Table I: Description of the samples studied in this research: density and mechanical properties by means of nanoindentation and micro-cantilever analysis

|  | Nanoindentation | | | | | | Micro-cantilevers | | | |
|---|---|---|---|---|---|---|---|---|---|---|
|  | Fiber | | | Matrix | | | | | | |
|  | | | | | | | Grain orientation | | | |
|  | | | | | | | parallel | | perpendicular | |
|  | H | Er | E | H | Er | E | $K_{Ic}$ | E | $K_{Ic}$ | E |
|  | GPa | | | | | | $MPa\sqrt{m}$ | GPa | $MPa\sqrt{m}$ | GPa |
| GA-1 | 26.44 ± 4.10 | 245.97 ± 29.03 | 303.0 ±.28.8 | 38.5 ± 2.6 | 333.5 ± 14.8 | 455.0 ± 14.5 | | | | |
| GA-2 | 30.00 ± 4.55 | 273.67 ± 23.47 | 347.8 ± 23.2 | 43.0 ± 2.4 | 347.4 ± 14.8 | 482.3 ± 14.5 | 3.57 ± 0.36 | 473.49 ± 15.23 | 2.39 ± 0.31 | 470.21 ±.19.35 |

Nanoindenation was performed on the samples, in both individual fibers and matrix, using a Micro Materials NanoTest™ (UK). The nanoindenter tip was calibrated using fused silica prior to the measurements. The calibration leads to an area function correcting for geometrical inconsistencies in the used diamond tip. The indents were placed 2 to 3 μm apart and had depths of 150nm or 200nm for higher spatial resolution. The nanoindentation experiment was carried out using the depth vs. load hysteresis with indents being terminated using depth control. The thermal drift correction data was collected post-indentation. The hardness (H) and reduced indentation modulus (Er) were calculated from the indents utilizing the Oliver and Pharr method [24].

A Quanta 3D field emission gun (FEG) scanning electron microscope with focused ion beam (SEM/FIB) has been used in this research. FIB surface milling provided a clean surface for better imaging of the samples and to prepare cross-section of the fibers. Energy dispersive spectroscopy was carried out on the fibers using an Oxford EDS attached to the SEM/FIB. A Renishaw Invia Laser Raman spectrometer was used for the characterization. The excitation wavelength used was

488 nm line of an Ar⁺ ion laser at an incident power of 10 mW. The area was analyzed using a 2 μm diameter spot through a standard ×50 microscope objective. The spectra were collected with a 10 s data point acquisition time, a spectral range of 200-2000 $cm^{-1}$ and a spectral resolution of 2 $cm^{-1}$. Image J [25] has been used for the image processing of the SEM images.

Fracture toughness has been studied using a Hysitron Pi-85 Picoindenter and a Micro Materials NanoTest™. Micro cantilevers were manufactured in grain direction (radial to the fiber) and perpendicular to the grain direction (tangential to the fiber). There were 11 (6 parallel and 5 perpendicular) micro cantilevers manufactured by the S.G. Roberts method [26]. The FIB procedure was as follows: three trenches were cut 20 μm wide and 10 μm deep at 15 nA beam current on three sides to shape bend bar. These trenches formed a U-shaped trench cut. The geometry was refined using 1 nA beam current. The sample was titled to 45 degrees along the length axis to shape cantilever and the base of the cantilever was undercut from both sides using a 3 nA beam current was used. Due to the anticipated fracture toughness measurements the bend bars were notched using a 10 pA beam current. The fracture toughness $K_{Ic}$ can be determined using the equation 1 from D. Di Maio and S.G. Roberts;

$$K_{Ic} = \sigma_c \sqrt{(\pi a)} \, F(a/b) \qquad (1)$$

where $\sigma_c$ is the fracture stress, a is the crack length (notch depth) and F(a/b) is a dimensionless shape factor dependent on sample geometry [26]. The stress is calculated using a standard mechanical approach.

## Results

### Mechanical properties by nanoindentation

The mechanical properties, H and Er, were evaluated in cross-section mode in order to measure the hardness of the fibers. In Figure 1a load-displacement curves for sample GA1 has been represented. 2D dimensional hardness maps of the material were gained and plotted in 3D with the hardness on the Z-axis as can be seen in Figure 1b. The post indentation analysis by SEM is shown in Figure 1c. The pyramidal imprints from indentations are easily identified and microcracking at the sites is observed.

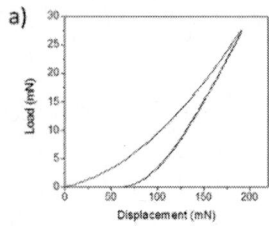 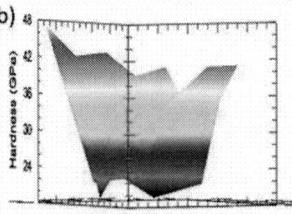 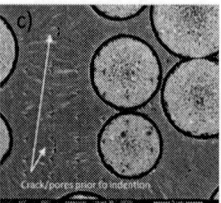

Figure 1.  (a)  Typical load displacement curves obtained from GA-1,
(b)  3D Plot of the hardness values on a particular fiber and
(c)  SEM image after indentation.

Since the location of each individual indent was located utilizing SEM all hardness data can be summarized in Table 1 and the words "fiber" or "matrix" have been used to indicate the location of the indents. On average, the fibers (26-30 GPa) are significantly softer than the matrixes (38-

43 GPa). Similar values for the matrix have been found in [27] for CVI produced SiC in Triso particles.

## Mechanical properties by micro cantilevers

During the SEM observation it was found that the SiC matrix grains are elongated along the radial direction from the fiber surface. In order to evaluate if there is any difference between the fracture toughness depending on the direction of grain growth, two different groups of bend bars were manufactured using the FIB instrument on sample GA-2. Two bend bars were cut parallel to the grain growth (radial direction to a fiber) while the two bend bars were cut perpendicular to the grain growth (tangential direction to the fiber). In Figure 2a, a load-displacement curve obtained from a parallel oriented grain bend bar has been plotted. Ultimate failure of this cantilever could be observed ~ 640 µN. The average values of all the micro cantilever fracture toughness tests are shown in Table I and in Figure 2b. The micro testing showed that the cantilevers in parallel with the grains had slightly higher fracture toughness than the micro cantilevers perpendicular to the grain orientation.

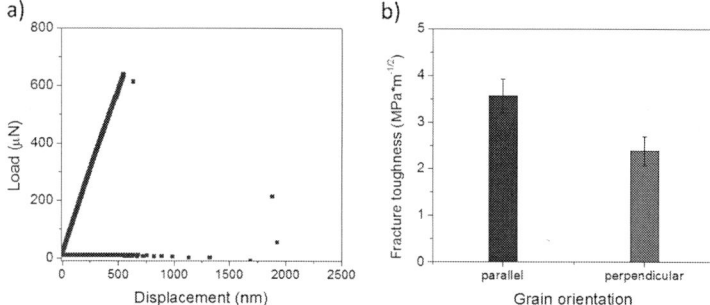

**Figure 2.** Load-displacement curve obtained from a parallel grain orientation cantilever (a) and calculated fracture toughness (b).

## Microstructural analysis

Microstructural analysis using SEM has been performed in order to understand the lower mechanical behavior of the fibers. As can be seen in Figure 3a for GA-2, the fiber appears to have a large amount of small black regions that appear significantly different than the surrounding SiC material. Energy dispersive spectroscopy (EDS) has been used to elucidate the elemental chemical composition of the black spots and the spectra for Si and C are shown. A gradual decrease of the Si signal could be observed across the center of the fiber, however, C signal increased in the middle and at the edges of the fiber. To further analyze these spots, micro Raman spectroscopy was conducted on the fibers.

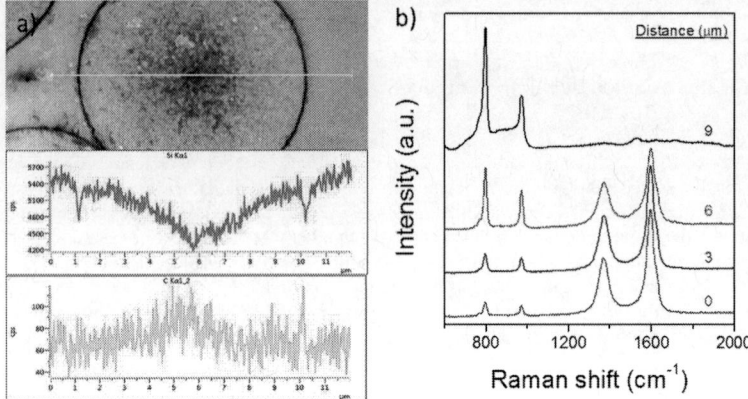

**Figure 3.** SEM top view and EDS line scan from one of the fibers and (a) and Raman spectra from one of the fibers in GA-2 (b).

In Figure 3b the Raman spectra from 200 to 2000 cm$^{-1}$ for the fiber and matrix from sample GA-2, which were taken from a radial distance from the center of the fiber, are shown. The distance from the center of the fiber has been labeled on the plot. A distance of 0 μm corresponds to the core-fiber, 3 μm to the inner-fiber, 6 μm to the outer-fiber and edge, and 9 μm to the matrix. Peaks have been identified corresponding to the different vibrational modes of SiC and C. β-SiC phase is denoted by the presence of two narrow peaks at around 796 and 972 cm$^{-1}$ and the graphitic carbon phase is characterized by the bands at 1370 and 1596 cm$^{-1}$ [28,29] A decrease in the intensity of the carbon signal and an increase of the SiC vibration modes are observed moving out from the core of the fiber to the matrix.

## Discussion

When analyzing the results of the nanoindentation performed on the SiC/SiC composite sample there is a noticeable difference in the H and $E_r$ between the matrix and fiber. The matrix has values of approximately 38 GPa in hardness meanwhile the fiber showed 26 GPa. The fiber is 25% softer than the matrix. This difference may just be an artifact of composite behavior on the microscale due to the possibility of: a) movement of the fiber during testing due to the pyrolytic carbon interface layer, similar to what occurs during fiber loading, or b) the pyrolytic carbon interface layer being soft enough to allow the SiC to expand during the indentation. However, it did not seem likely that the fiber was sliding since they were not depressed into the matrix after testing. It also did not seem likely the fiber was expanding during the indentation process because the carbon interface was not displaced after testing.

It is also possible that the lower hardness is real, and is a result of inhomogeneous regions in the fiber that contain a higher amount of carbon as shown by EDS and Raman spectroscopy. The SEM image of the samples indicates darken spots in the fibers themselves (Figure 4a). Figure 4 also displays a gradient throughout the fiber with high concentration of dark dots in the center and lower concentration outside while the fibers grain structure is visible. It appears that the dark dots are located at the grain boundaries of the fibers grains but more detailed TEM or high resolution SEM investigation would be needed to verify this.

From the SEM observation alone, it is difficult to determine if these dark spots are either carbon-rich regions or internal porosity. However, EDS and Raman spectroscopy allowed verifying that the dark spots are carbon enrichments areas and mainly carbon. The graphite material is believed to be creating the significantly weaker mechanical properties than those of the SiC matrix. In order to quantify the amount of graphite and its effect towards mechanical properties Image J analysis was performed and it is shown in Figure 4b. Since it is assumed at this point that the dark dots are either voids or low Z-number material it is safe to assume that the mechanical properties are significant weaker than that of SiC matrix. It was found that in the center of the fiber 30-35% area consist of graphite while some individual fibers can be as high as 50%. Applying the simple rule of mixtures allows the estimation the expected changed mechanical property of the fiber due to the graphite particles. This approach is expressed in Equation 3.

$$H_{estimated} = Vol\%_{graphite} * H_{graphite} + Vol\%_{SiC} * H_{SiC} \quad (2)$$

Assuming that 70% by volume in the center of the fiber is SiC and its hardness is 38GPa and 30% is Graphite and its hardness is 4 GPa, as previously by Mann et al in similar fibers [30]. it can be calculated that the hardness of the center of the fiber is approximately 27.8 GPa which is in the range that was measured with the nanoindentation (26 GPa). The same calculation can be made for the Young's modulus (E) which results of 327 GPa. Both theoretically calculated values for H and E are in reasonable agreement with the actual measured values considering the error of the measurements. This approach ignores the fact that the Graphite-SiC also has interface strengthening.

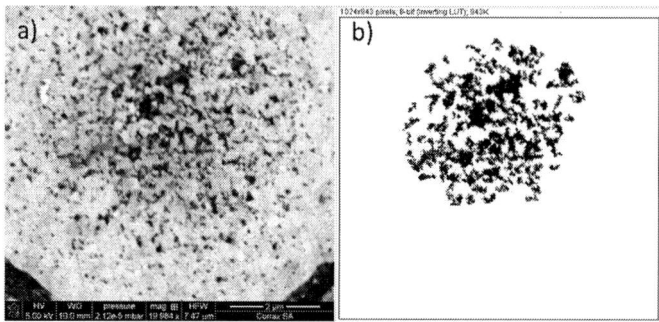

Figure 4.    SEM image top-view of GA-1 (a) and Image j analysis of the center of the fiber (b).

The microstructure of the SiC/SiC composite reveals that when the SiC in the matrix begins to crystallize the grains grow radially outward from the fiber. This phenomenon can be seen in Figure 1c. As the grains grow outward from the fiber and meet other grains prevalent voids and cracks are produced which could be seen in Figure 1c. These zones could be the initiators for cracks in the matrix and have lower H and Er values. The values for the void zone are approximately 34.2 GPa for the hardness and 305.5 for the Young's modulus when the matrix hardness value was 40 GPa. Further experiments were performed to see if there was a hardness gradient in the matrix due to the radial formation pattern of the grains. When the nanoindentation testing was completed the results indicated that there was no hardness gradient in the matrix.
Another structure that was observed is composed of the small grains adjacent to the fiber that form a ring around the fiber. We refer to this external ring as the helo-structure. The outer edge

of these rings are possible initiator of cracks, they are visible in Figure 1c under the top fiber. Cracks appeared after indentation much more frequently near the helo-rings than in other areas leading us to hypothesize that the outer edge of these rings are potential initiators of cracks.

The strain of each micro cantilever test was measure and used to make stress vs. strain curve which could be used to calculate the Young's modulus of the material and compare with the nanoindentation performed earlier. The strain was found using the formula below.

$$\varepsilon(x) = \frac{(c-x)z*6*w(x)}{c^3} \qquad (3)$$

Where c is overall length of cantilever, x is the lever length (point of contact to the notch), z is half the height, w(x) is the displacement. This formula allowed calculation of the strain at the point of contact of the load. A representative plot of the stress vs. strains curves is shown below in Figure 5. A linear fit has been carried out and it has been included in the plot. A Young's modulus of ~472 GPa has been obtained from this bend bar.

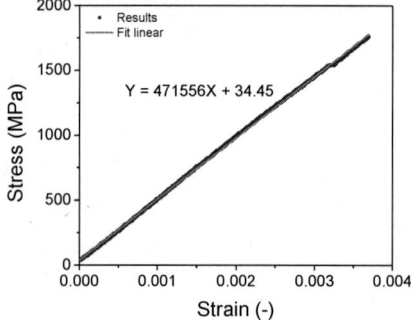

**Figure 5.** Representative stress vs. strain plot for a micro cantilever (parallel orientation).

The micro cantilever test gave the fracture toughness of matrix SiC of 3.47 ± 0.36 MPa m$^{1/2}$ for the parallel and 2.29 ± 0.31 MPa m$^{1/2}$ for the perpendicular. The average Young's modulus calculated using the micro cantilevers were 473.5 and 470.2 GPa for the parallel and perpendicular grain orientation for the matrix of the hardest sample. The value obtained by nanoindentation was 482.3 GPa. The values of the Young's modulus calculated with the micro cantilever methods are very similar to the nanoindentation indicating that the sample does not have a disadvantage in case of being used in any direction. However, more investigation is necessary to correlate the microstructure fracture toughness measurement to a macrofacture toughness measurement.

**Conclusion**

The work performed here clearly shows that nanoindentation can be used to test the individual components of a composite and evaluate its mechanical properties. The ability to see each component properties individually allows us to see how they interact as whole composite and identify potential weak spots. The nanoindentation results indicated that the fiber is significantly softer than the matrix. EDS and Raman confirmed nanometer sized graphite particulates embedded in the fiber. After the study of the phases, the rule of mixture has been used successfully to verify the H and E of the fibers. The results of the micro cantilever carried out on the hardest sample showed that the grain orientation in the matrix had a small but no significant effect of the fracture toughness of the matrix. These results give modelers more information about the composite and if incorporated into the modeling, can lead to more accurate predictions of the overall composites behavior in a structural application. It could also lead to a better understanding of the mechanical property changes that happen during the lifetime of the composite because small scale testing can observe the changes in each component of the composite. These observations of the changes in each component would lead to better design of the composites.

# References

[1]  The stationary office "our energy future- creating a low carbon economy" Energy White Paper, UK, (2003).

[2]  Report of National Energy Policy Development Group "Reliable, Affordable and Environmentally Sound Energy for American's Future" National Energy Strategy, US/DOE, May (2001).

[3]  S. Kubo, H. Nakajima, S. Kasahara, S. Higashi, T. Masaki, H. Abe, K. Onuki. "A demonstration study on a closed-cycle hydrogen production by the thermochemical water-splitting iodine–sulfur process." Nucl. Eng. Des. 233 (2004)

[4]  N. Woudstra, , T. Woudstra, A. Pirone, T. Stelt "Thermodynamic evaluation of combined cycle plants" Ener. Con. Manag. 51

[5]  A.M. Bassily "Modeling, numerical optimization, and irreversibility reduction of a triple-pressure reheat combined cycle" Ener. 32 (2007)

[6]  Y. Katoh, D.F. Wilson, C.W. Forsberg (contributors) Lance L. Snead, David F. Williams, "Assessment of Silicon Carbide Composites for Advanced Salt-Cooled Reactors" ONRL/TM-2007/168, (2007)

[7]  C.P. Deck, H.E. Khalifa, B. Sammuli, T. Hilsabeck, C.A. Back, "Fabrication of SiC-SiC composites for fuel cladding in advanced reactor designs", Progress in Nuclear Energy, v.57 pp.38-45 (2012)

[8]  R.J. Price, G.R. Hopkins, "Flexural Strength of Proof-Tested and Neutron-Irradiated Silicon Carbide," J. Nucl. Mater. 732 (1982)

[9]  Naslain R. : Bunk WGJ, editor." Thermostructural ceramic matrix composites: an overview, advanced structural and functional materials". Berlin: Springer±Verlag, (1991)

[10]  R. J. Price, "Properties of silicon carbide for nuclear fuel particle coatings" Nuc. Technol. 35 (1977)

[11]  Marshall DB, Evans AG. "Failure mechanisms in ceramic-Fiber/ceramic-matrix composites" J. Am. Ceram. Soc. (1985)

[12]  S. Zhu, M. Mizuno, Y. Kagawa, Y. Mutoh, "Monotonic tension, Fatigue and creep behavior of SiC-fiber-reinforced SiC-matrix composites: a review" Comp. Sci. and Tech. 59 (1999)

[13]  A. Kohyama, "Advanced SiC/SiC Composite Materials for Fourth Generation Gas Cooled Fast Reactor*," Key Eng. Mater. (2005)

[14]  S. Nogami, N. Otake, A. Hasegawa, Y. Katoh, A. Yoshikawa, M. Satou, Y. Oya, K. Pkuno, "Oxidation behavior of SiC/SiC composites for helium cooled solid breeder blanket" Fus. Eng. And Des. 83 (2008)

[15] B.A. Pint, J.L. Moser, P.F. Tortorelli "Investigation of Pb-Li compatibility issues for dual coolant blanket concept" J. Nucl. Mater. 367–370 (2007)

[16] T. Nozawa, T. Hinoki, A. Hasegawa, A. Kohyama, Y. Katoh, L.L. Snead, C.H. Henager Jr., J.B.J. Hegeman, " Recent advances and issues in development of silicon carbide composites for fusion applications," J. Nuc. Mater. 622-627 (2009)

[17] Evans AG. "Perspective on the development of high-toughness ceramics" J. Am. Ceram. Soc. 1990.

[18] M. Ferraris, V. Casalegno, S. Rizzo, M. Salvo, T.O. Van Staveren, J. Matejicek "Effects of neutron irradiation on glass ceramics as pressure-les joining materials for SiC based components for Nuclear Applications" Jour. Nuc. Mat. 429 (2012)

[19] T. Hinoki, E. Lara-curzio and L.Snead, "Mechanical Properties of High Purity SiC Fiber-Reinforced CVI-SiC Matrix Composites," Fus. Sci. and Tech., 44 ( 2003)

[20] L.L. Snead, R.H. Jones, A. Kohyama, P. Fenici "Status of silicon carbide composites for fusion," J. Nucl. Mater., 26 (1996)

[21] Effinger, M.R.& Genge, G.G.& Kiser, J.D., "Ceramic composite turbine disks for rocket engines", Adv. Mater. Process. 157, 2000

[22] T. Noda, M. Fujita, H. Araki and A. Kohyama "Impurities and Evaluation of Induced Activity of CVI SiCf/SiC composites," Fus. Eng. and Des., 51-52, 99 (2000)

[23] T.Ishikawa "Advances in Inorganic Fibers," Adv. Polym. Sci. (2005)

[24] W.C. Oliver and G. M. Pharr "An improved technique for determining hardness and elastic modulus using load and displacement sensing indentation experiments" J. Mater. Res., 7 (1992)

[25] http://rsb.info.nih.gov/ij

[26] D. Di Maio, S.G. Roberts, "Measuring fracture toughness of coating suing focused-ion-beam-machined microbeams"J. Mater. Res., Vol. 20, (2005)

[27] P. Hosemann, J. N. Martos, D. Frazer, G. Vasudevamurthy, T. S. Byun, J. D. Hunn, B. C. Jolly, K. Terrani, M. Okuniewski, "Mechanical characteristics of SiC coating layer in TRISO fuel particles" Journ. Of Nuc. Mat. 442 (2013)

[28] T. S. Perova, R. A. Moore, K. Berreth, K. Maile, A. Lyutovich. "MicroRaman spectroscopy of protective coatings deposited onto C/C–SiC composites" Mat. Sci & Technol 23 (11) (2007) 1300-1304

[29] Ph. Colomban, G. Gouadec, L. Mazerolles. "Raman analysis of materials ... the example of SiC fibers" Mater. Corros. 53 (2002) 306-315

[30] A.B. Mann, M. Balooch, J.H. Kinney, and T.P. Weihs. "Radial variations in modulus and hardness in SCS-6 silicon carbide fibers" J. Am. Ceram. Soc., 82 (1) 111-116 (1999)

# PROCESSING FRACTURE TOUGHNESS AND DAMAGE MECHANICS STUDIES OF MMC FOR AEROSPACE

Ajit Bhandakkar[1], R C Prasad[1] and Shankar M L Sastry[2]

[1]Department of Metallurgical Engineering and Materials Science,
**Indian Institute of Technology** (IIT) Bombay
Bombay, **India**

[2]Mechanical, Aerospace and Structural Engineering
**Washington University** in St. Louis
St. Louis, MO, **USA**

## Abstract

Aluminum metal matrix composites (MMCs) find important applications in aerospace where specific stiffness is important. Low cost fly ash reinforcement is widely used in aluminium metal matrix composite due to its property of low density, high young modulus and strength apart from good mechanical and chemical compatibility & thermal stability. In the present paper the influences of weight fraction of particulate reinforcement on tensile, fracture toughness have been evaluated. The fracture toughness ($K_{IC}$) of the aluminum fly ash composite increases by almost 24% that of base alloy. The load and COD plot shows hysteresis loop in loading and unloading compliance curve. This is indicative of crack closure. The reason for crack closure may be surface roughness resulting from reinforcement particles in the composites. The fracture behavior and micro-mechanism of failure in base alloy and composites have been observed under SEM and optical microscopy.

**Keywords**: Fracture Toughness, Aluminium Fly Ash Composites, MMCs, Damage Mechanics.

## Introduction

The need for lightweight, high strength low cost materials is the great challenge even today when the global oil resources is on decline and increase in the fuel efficiency of engines has become highly desirable. As the strength and stiffness of a material increases, the dimensions, and consequently, the mass, of the material required for a certain load bearing application is reduced. This leads to several advantages in the case of aircraft and automobiles such as increase in payload and improvement of the fuel efficiency [1]. Monolithic materials often find limitations with increasing performance requirements in various environments, hence there has been need to find new materials with tailor made properties .The main focus now shifted to Aluminum alloy metal matrix composites because of its unique combination of good corrosion resistance and excellent mechanical properties. However in recent times fly ash which is by product of coal burning is drawing lot of attention as reinforcement for MMCs due to its low cost and reduction in environmental pollution. The ash particles, generally being hollow in nature, display lower densities while oxides present as constituents make them possess high modulus and strength thereby enhancing specific strength and stiffness along with lower densities compared to many metal based systems.

In this paper, fabrication of 6061 Al/ fly ash MMC by liquid metallurgy route with secondary processing hot extrusion is discussed. The Aim of the present study is to study the mechanical properties, fracture behavior and micro-mechanisms of failure in AA6061 aluminum unreinforced alloy and metal matrix composites.

## Experimental

### Material

Aluminum alloy AA 6061 is used as base matrix with composition (weight percent) listed in Table.1 The reinforcement used is fly ash grade P60 having particles of sizes 25-45 in 5%, and 10% by weight and the chemical composition of fly ash reinforcement is as per Table.2

Table 1.     Chemical composition of matrix alloy AA6061

| Grade   |      | % Elements |      |      |      |      |       |      |
|---------|------|------------|------|------|------|------|-------|------|
| AA6061  | Al   | Cu         | Mn   | Mg   | Zn   | Fe   | Cr    | Si   |
|         | Base | 0.23       | 0.21 | 0.86 | 0.057| 0.33 | 0.094 | 0.63 |

Table 2.    Chemical composition of fly ash reinforcement

| Grade | % Elements | | | | | |
|---|---|---|---|---|---|---|
| Indian Fly ash P60 | $Al_2O_3$, $SiO_2$, $Fe_2O_3$ | CaO | MgO | $Na_2O$ | $K_2O$ | $SO_3$ |
|  | 92.49 | - | 2.13 | 0.73 | - | 1.06 |

## Processing of Aluminum Fly ash composites

Fig.1 and Fig.2 shows the experimental setup for fabrication of aluminum fly ash composite through liquid metallurgy route. About 1 kilograms of the AA 6061 alloy is cleaned and loaded in the silicon carbide crucible and heated to above its liquidus temperature. The temperature was recorded using chromel-alumel thermocouple. To maintain the solid fraction of about 0.4, the temperature of the melt was lowered before stirring. The specially designed mechanical graphite stirrer is introduced into the melt and stirred at ~ 400 rpm as shown in **Fig.3**. The depth to which the impeller was immersed is approx $1/3^{rd}$ the heights of the molten melt from the bottom of the crucible. The preheated ($800^0C$) fly ash particulates (25-45μm) were added through a preheated pipe by manual tapping into the slurry, while it was being stirred. **Table 3** gives the stir casting process details. A post-addition stirring time of 30 min was allowed to enhance the wetting of particulates by the metal. The temperature of the slurry was sufficiently raised above the melting range of the matrix alloy before pouring the composite melt into preheated permanent mould.

Table 3.    Stir casting process details for fabrication of aluminum fly ash composites

| Sr.No | Composite System | Fly ash Size (μm) | Preheat Temp of fly ash reinforcement | Total Stirring time | Pouring Temp. ($^0C$) |
|---|---|---|---|---|---|
| 1 | 6061+5% Fly ash (wt%) | 25-45 | $800^0C$ | 30min | 750 |
| 2 | 6061+10% Fly ash (wt%) | 25-45 | $800^0C$ | 30min | 800 |

Figure 1.   Experimental set up for processing of AA6061 / fly ashomposites

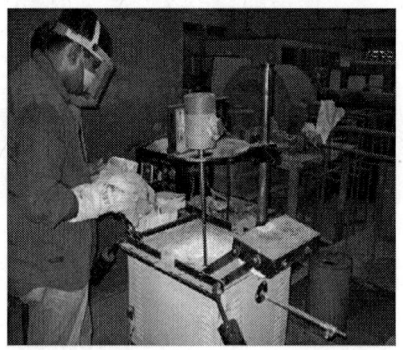

Figure 2.   Experimental set up for fabrication of composite

**Figure 3.** Graphite stirrer for uniform distribution of Aluminum fly ash composite

### Secondary processing

The as-cast composite billets were extruded for (64 % reduction) at 450°C (Soaking for 4 hrs) in order to get rid of the porosities induced during primary processing. It also improves the distribution of the reinforcement in the matrix. Secondary processing improves distribution of fly ash reinforcement in the matrix, imparts directional properties, whereby mechanical properties are improved. The hot extrusion details of Metal Matrix Composite (AA6061 + fly ash ) are shown in Table 4.

**Table 4.** Extusion ratio used for secondary processing of AA6061/fly ash composites

| Material | Initial Diameter(mm) | Final Diameter | % Reduction |
|---|---|---|---|
| AA 6061 base alloy | 49.5 | 17.74 | 64.16 |
| AA6061 +5%P60 | 49.5 | 17.74 | 64.16 |
| AA6061 + 10%P60 | 49.5 | 17.74 | 64.16 |

### Specimen Preparation

The tensile specimens were fabricated from the extruded rods of the base metal and composite extrusions. as shown in **Fig.4** as per ASTM E8 were used for tensile testing. The SENB specimens for $K_{IC}$ are prepared in LT direction with notch and intended direction perpendicular to the rolling direction as per ASTM E-1820 and ASTM E-647 standards as shown in **Fig 5 and Fig.6**

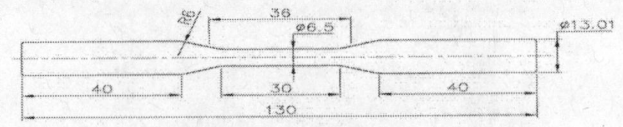

Figure 4.     Tensile test specimen as per ASTM E-8

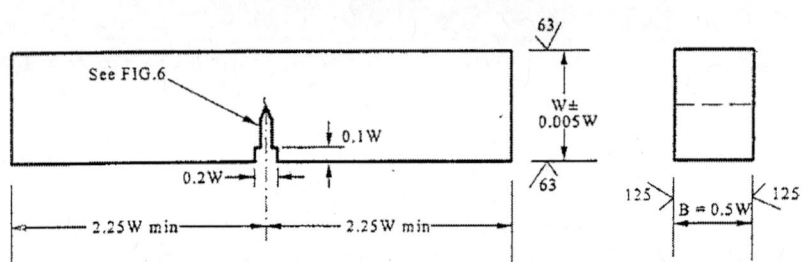

Figure 5.     Fracture toughness test specimen

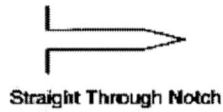

Straight Through Notch

Figure 6.     Fatigue crack starter notch configuration

## Results and Discussion

### Fracture toughness Testing

Plane strain fracture toughness ($K_{IC}$) tests were conducted on BiSS 50 KN servo hydraulic Universal Testing Machine, using SENB as per ASTM E-1820. The conditional fracture toughness was calculated using following Eqn.1 and the fracture toughness of base alloy and composites listed in Table.5

### Equation 1 Fracture Toughness of composite $K_Q$

$$K_Q = \frac{P_Q}{B.W^{\frac{1}{2}}} xLxf\left(\frac{a}{w}\right),$$

$K_Q$ = Conditional Fracture Toughness
$P_Q$ = Load value obtained by 95% secant line.
L = Span length
A = Crack length
W = Width of the specimen

Table 5.　　Fracture toughness of AA6061/fly ash composites

| Grade | Condition | Thickness (in) | Span length | a/w | Peak Load (lbs) | Fracture Toughness in MPa√m |
|---|---|---|---|---|---|---|
| 6061 base | Extruded | 0.25 | 0.50 | 0.50 | 1195 | 28.09 |
| 6061-5% FA | Extruded | 0.25 | 0.50 | 0.50 | 1190 | 31.61 |
| 6061-10% FA | Extruded | 0.25 | 0.50 | 0.50 | 1182 | 35.11 |

## Mechanical Properties

### Tensile Properties

The mechanical properties such as ultimate tensile strength, 0.2% yield strength and percentage elongation have been evaluated for AA 6061 base alloy and fly ash composites and are listed in Table.6 and shown in Fig.7. The hardness of the base alloy and composite listed in Table.7

Table 6.　　Tensile test results of AA6061 base alloy and AA6061/fly ash composites

| Condition (Rolled) | 0.2% Y.S (MPa) | UTS (MPa) | % Elongation |
|---|---|---|---|
| AA6061 + 0% wt. fly ash | 122 | 184 | 10 |
| AA 6061 + 5% wt.fly ash | 136 | 222 | 3.45 |
| AA 6061 + 10% wt. fly ash | 184 | 249 | 3.20 |

## Hardness

The hardness have been evaluated for AA 6061 base alloy and fly ash composites with Leco Vickers Micro hardness tester and the hardness values of the base alloy and composites are listed in Table.7. The hardness of the aluminum fly ash composites increase with the fly ash reinforcement. The SEM micrograph in Fig.7a shows uniform distribution of the fly ash reinforcement in the aluminum matrix. The increase in the micro hardness is due strain fields created around fly ash particles because of the difference in the thermal expansion coefficients of aluminum base alloy and fly ash particles. The strain fields' piles up dislocations and the interaction between dislocations and fly ash particles offer resistance to the propagation of cracks. The grain refinement provided by the fly ash particles during solidification is also responsible for increase in the micro hardness.

Table 7.     Hardness of AA6061 base alloy and fly ash composites

| Grade | Hardness ( VPN) |
|---|---|
| AA6061   base alloy | 48 |
| AA 6061 + 5% Fly ash | 52 |
| AA 6061 + 10% Fly ash | 61 |
| AA6061  + 20% Fly ash | 70 |

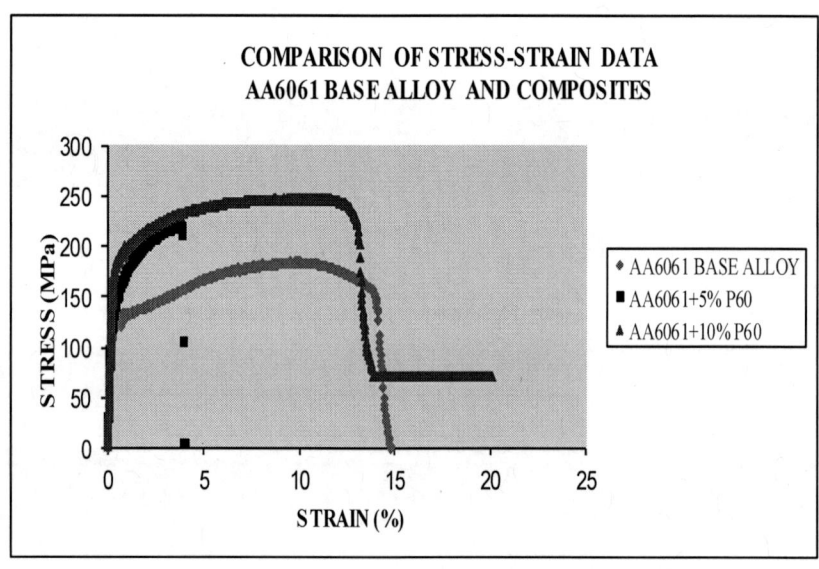

Figure 7.     True Stress True Strain plot of AA6061 base alloy and composites

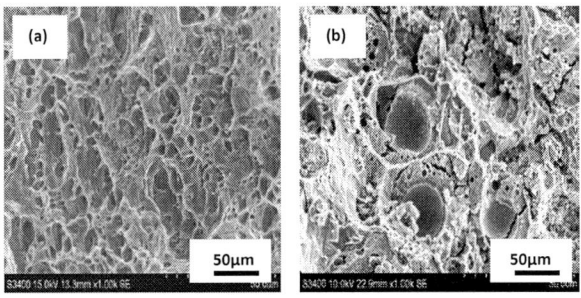

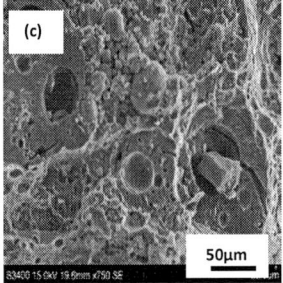

Figure 7.   Fractographs showing the
(a)   AA6061 base alloy, and
(b)   AA6061-5% FA composites
(c)   AA6061-10% FA composites

Tensile test results as listed in Table.6 and Figure.7 of aluminum fly ash composites at room temperature indicates that with the increase in the fly ash reinforcement from 0% to 10% the yield strength increases from 122 MPa to 184 MPa and the tensile strength increases from 184 MPa to 249 MPa . This increase in the yield strength and tensile strength may be attributed due to the presence of high dislocation densities at the particle/matrix interface is due to the difference in coefficient of thermal expansion between aluminium alloy matrix and fly ash particles. The tensile fracture surface was observed under SEM to study the modes of fracture as shown in Fig.12 and Fig.13.The large dimples are associated with the reinforcing particles and small dimples are associated with the fine second phase particles in the matrix. The dimples observed in the aluminum fly ash composites were shallow compared to that observed in unreinforced alloy. The presence of particles in the ductile aluminum alloy matrix promotes low plasticity and condition of high stress triaxiality and voids nucleate at low strain and coalescence is dominated only by their transverse growth. This prevents the extensive growth of voids results in shallow dimples in composites alloy as compared to base alloy.

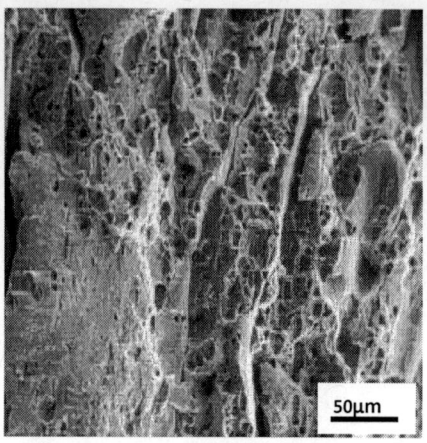

Figure 8.    AA6061 base alloy having mixed mode type tensile fracture

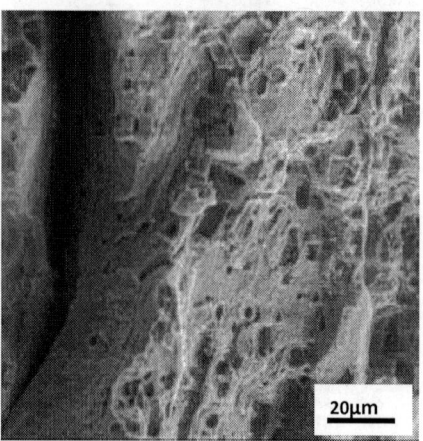

Figure 9.    Lots of cracks are present with river line patterns

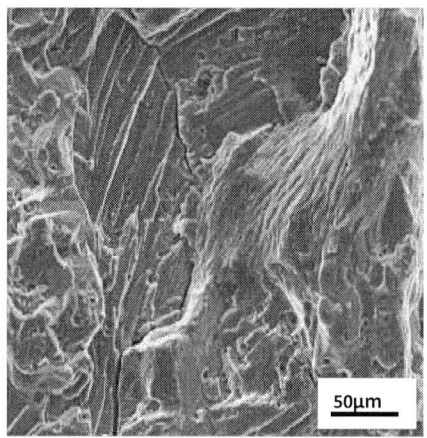

Figure 10. AA6061 - 5%fly ash composite showing dimples

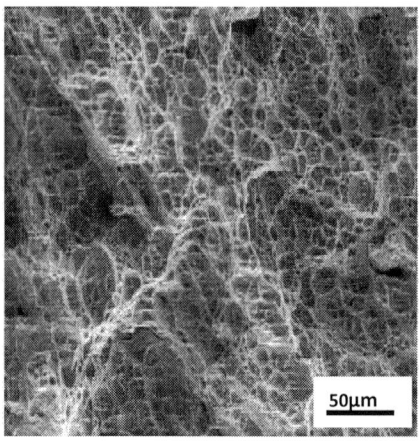

Figure 11. AA6061 -10 %fly ash composite showing dimples

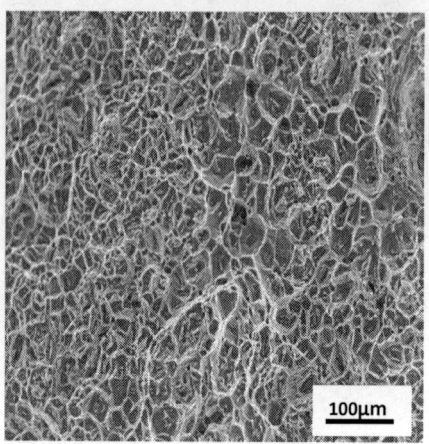

**Figure 12** AA6061-5% fly ash composite shows dimples

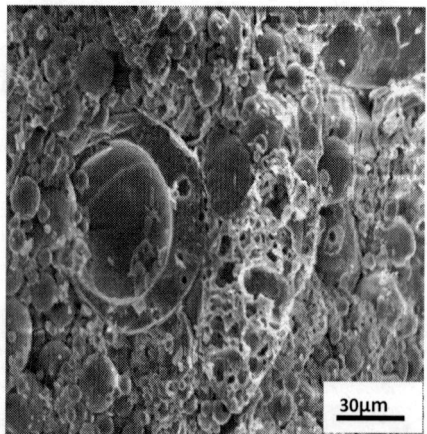

**Figure 13** AA6061-10% fly ash composite showing fracture mode, interface debonding and particle fracture

The fracture toughness ($K_Q$) of AA6061 base alloy and AA6061fly ash composites is varied between 31-35 MPa$\sqrt{m}$ as compared to 28 MPa$\sqrt{m}$ for the unreinforced alloy. The typical fracture surface of the three point bend test specimen of composites shows bimodal dimples as shown in Fig.8. These dimples are characteristics of failure occurring by void nucleation, growth through the matrix and coalescence. The presences of reinforcement particles restrict the deformation of the matrix thereby reducing the process zone considerably. In addition the particle matrix deboning was also observed as shown in Fig.9.

## Conclusions

1. Uniform distribution of fly ash particles in the aluminum matrix was obtained by liquid metallurgy route of stir casting followed by hot extrusion.

2. The yield strength, Tensile strength of AA6061 fly ash metal matrix composites increases with the increase in fly ash reinforcement, however % elongation decreases with the increase in fly ash reinforcement.

3. The plane strain fracture toughness $K_{IC}$ of the aluminum fly ash composites is improved with the addition of fly ash reinforcement and therefore is having high crack arrest capability. They may therefore, be considered as potential candidate materials for aerospace sectors.

# References

1. S. R. Bakshi, D. Lahiri and A. Agarwal*,"Carbon nanotube reinforced metal matrix composites – a review " International Materials Reviews 2010 VOL 55 NO 1

2. VK Lindroos & MJ Talvitie, "Recent advances in metal matrix composites", *Journal of Materials Processing Technology*, 53(1995) pp 273-284

3. EA Starke Jr & JT Staley, " Application of modern aluminium alloys to aircraft", *Prog. Aerospace Science*, Vol.32, pp 131-172, 1996

4. KA Lucas & H.clarke, "Corrosion of Aluminium-based Metal Matrix Composites", Research Studies Press Ltd, 1993

5. ND Alexopoulos & P Papanikos, "Experimental and theoretical studies of corrosion-induced mechanical properties degradation of aircraft 2024 aluminium alloy", *Materials Science and Engineering* A, 498(2008), pp 248-257

6. J-P Immarigeon et al, "Lightwight materials for aircraft applications", *Materials Characterisation,* 35(1995), pp 41-67

7. I.J.Polmear, "Light alloys : From traditional alloys to nanocrystals", Elsevier, Fourth Edition, 2006, pp 75, 117-141

8. Emenike Raymond Obi, "Corrosion Behaviour of Fly Ash-Reinforced Aluminum-magnesium Alloy A535 Composites" Master of Science Thesis, University of Sasketchewan, Saskatoon, Sep 2008.

9. FL Matthews and RD Rawlings, "Composite Materials: Engineering and Science", I Edition, Chapman & Hall, 1994

10. Krishan K.Chawla, *"Composite Materials: Science and Engineering"* II Edition, Springer, 1998

11. TPD Rajan et al, "Fabrication and characterization of Al-7Si-0.35Mg/Fly ash metal matrix composites processed by different stir casting routes", *Composites Science and Technology* 67 (2007) pp 3369-3377

12. S Sarkar, S Sen and Sc Mishra,"Studies on Aluminium-Fly ash Composite produced by Impeller Mixing", *Jounal of Reinforced Plastics and composites*, Vol 00, No.00/2008

13. H.Abdizadeh et al, "Comparing the effect of processing temperature on microstructure and mechanical behaviour of ZrSiO4 / TiB2 –Aluminium Composites", *Materials Science and Engineering* A, V 498 (2008), pp 53-58

14. Emmanuel Gikunoo, "Effect of fly ash particles on mechanical properties and microstructure of aluminium casting alloy A535", MS Thesis, University of Sasketchewan, Ottawa, Canada, Nov 2004

15. PK Rohatgi et al, "Pressure infiltration technique for synthesis of aluminium-fly ash particulate composite" *Materials Science and Engineering*, V 244(1998), pp 22-30

16. Boq-Kong Hwu, Su-Jien Lin and Min-Ten Jahn, "Effects of Process parameters on the properties of Squeeze-case SiCp-6061 Al Metal Matrix composites", *Materials Science and Engineering* A, 207(1996), pp 135-141

17. Jerzy Sobczak et al, "The technological aspects of ALFA Composites Synthesis", Solidification Processing of Metal Matrix Composites – Rohatgi Honorary Symposium, Edited by N. Gupta, and W.H. Hunt TMS (The Minerals, Metals & Materials Society), 2006

18. V Amigo et al, " Microstructure and mechanical behaviour of 6061 Al reinforced with silicon nitride particles, processed by powder metallurgy", *Scripta Materialia,* 42(2000), pp 383-388

19. H Siddhi Jailani et al, " Multi-response optimization of sintering parameters of Al-Si alloy/fly ash composite using Taguchi method and grey relational analysis", *International Journal of advanced Manufacturing Technology*, 2009

20. RQ Guo & PK Rohatgi, "Chemical reaction between aluminium and fly ash during synthesis and reheating of Al-fly ash composite", *Metallurgical and materials transactions B,* Vol. 29B, June 1998, pp 519-525

# INTERFACE AND BONDING OF COMPOSITE SYSTEMS

Advanced Composites for Aerospace, Marine, and Land Applications
Edited by: Tomoko Sano, T.S. Srivatsan, and Michael W. Peretti
TMS (The Minerals, Metals & Materials Society), 2014

# MULTI-SCALE MODELING OF CONTINUOUS CERAMIC FIBER REINFORCED ALUMINUM MATRIX COMPOSITES

Brandon A. McWilliams

Oak Ridge Institute for Science and Education
Aberdeen Proving Ground
MD 21005, USA

Chian-Fong Yen

U.S. Army Research Laboratory
Aberdeen Proving Ground
MD 21005, USA

## Abstract

Continuous ceramic fiber reinforced metal matrix composites (MMCs) have demonstrated very high specific strengths and stiffness but use in structural applications has been limited by their inherently low ductility. In this work, a multi-scale micromechanics based finite element approach is used to predict, and understand the effect of microstructure on the macroscopic behavior and failure mechanisms of continuous ceramic fiber reinforced MMCs. A hierarchal approach is proposed in which a micro-scale model is used to predict yarn level response based on the properties of the matrix and individual fibers. The yarn properties determined from the smaller scale model can then used as input in a micromechanical fabric composite model consisting of yarns and matrix to predict the continuum response of the MMC. Cohesive zones are used to model the interfacial properties at both scales. The effects of interfacial properties and defects on the continuum response are investigated. Simulation predictions are compared with experimental data.

## 1. Introduction

Due to their high specific strength and stiffness, continuous ceramic fiber reinforced metal matrix composites (MMCs) have potential as components in mass efficient structural applications. Successful integration of these materials requires the development of advanced multi-scale numerical models which incorporate continuum mechanics and micromechanics based techniques to understand the effects of microstructure and processing on the overall material performance under multi-axial loading at high strain rates. Continuous fiber MMC materials have traditionally exhibited low ductility and low toughness (brittle behavior) which has mostly limited their use to applications requiring high temperature and/or improved wear resistance compared to the unreinforced base alloys [1-2]. Micromechanical modeling of fiber MMCs is therefore critical to understanding basic deformation and failure mechanisms of these materials in order to fully optimize microstructure and processing to increase their ductility and toughness sufficiently for their use in structural applications.

In this paper we present the numerical simulation of a unidirectional continuous ceramic fiber reinforced aluminum metal matrix composite. This effort is considered as one piece within a larger multi-scale modeling framework aimed at modeling the hierarchical response of woven ceramic fabric metal matrix composite (Fig. 1). At the continuum scale the overall woven composite consists of yarns embedded in the metal matrix. At the next lower scale, the yarns are composed of many fibers embedded in a metal matrix. At both these scales the interface behavior plays a crucial role in the mechanical response in addition to the mechanical behavior of the reinforcement and matrix constituent phases. Understanding the deformation and failure mechanisms at both these scales is critical to be able to accurately predict the overall continuum response of the material. This work considers a unit cell of ceramic fibers embedded in metal matrix at the sub-yarn scale with relevant progressive damage and failure mechanisms included. These failure mechanisms include, fiber fracture, ductile matrix failure, and interface separation/decohesion. The results of this work can then be used within this proposed multi-scale modeling framework as the properties of a MMC yarn at the fabric scale. Experimental testing is used to characterize damage and failure mechanisms and as validation of the numerical results for transverse (to fiber axis) and normal loading.

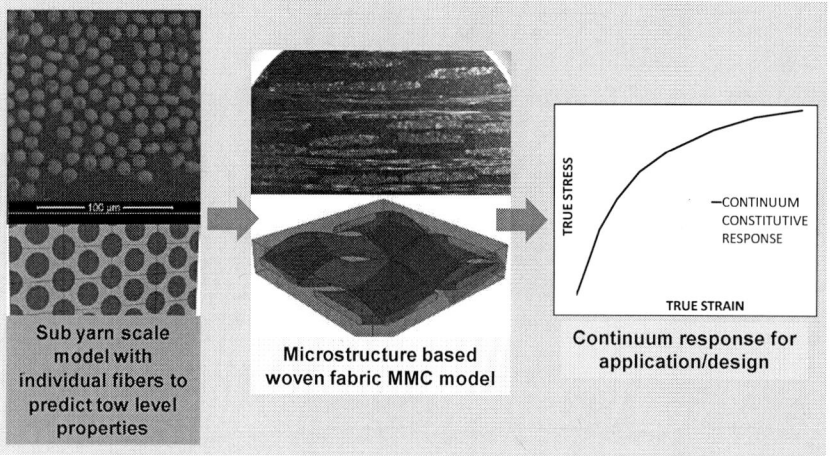

**Figure 1.** Conceptual schematic of hierarchal multi-scale modeling framework proposed for predicting the continuum scale deformation response of metal matrix composites including all relevant progressive damage and failure mechanisms at each length scale from micro to macro.

## 2. Methods

### 2.1 Materials

The ceramic reinforcement material used in this study were continuous alumina ($Al_2O_3$) fibers. The fibers were provided by 3M (St. Paul, MN, USA) and have the commercial designation Nextel 610®. The Nextel fibers are 12 μm in diameter and have a density of 3.9 g/cm$^3$. The fibers infiltration cast with an aluminum – 2% copper (Al-2%Cu) alloy by CPS Technologies (CPS, Norton, MA, USA). The nominal volume fraction of fibers in the final metal matrix composite was ~55% [3].

### 2.2 Numerical methods

A 3D periodic cell finite element model (FEM) was used to model the deformation response of an MMC with an assumed hexagonal packing of fibers. The response of the composite was modeled under uniaxial loading conditions in the axial (along fiber axis) and transverse (loading perpendicular to fiber axis) directions (Fig. 2). The model with periodic boundary conditions was implemented in the commercial software package ABAQUS/Explicit. The fibers were modeled as elastic with a Young's modulus of 380 GPa and Poisson's ratio of 0.23 [4-5]. The average strength of 2.97 GPa obtained experimentally [3] was used as the failure criteria to initiate damage in the fibers. The Al-2%Cu matrix was modeled as elastic-plastic with an equivalent plastic strain failure criterion. A Young's modulus of 70.9 GPa, Poisson's ratio of 0.3, and density of 2.8

g/cm³ were used for the elastic properties. The plastic constitutive behavior of the Al-2%Cu matrix was input in tabular form directly from the average stress-strain curve obtained from uniaxial tension experiments (Fig. 3). The strain to failure was taken as the average value obtained from uniaxial tension tests and a value of 0.127 was used.

**Figure 2.**   3D periodic model used in present simulations for transverse (A) and axial fiber direction (B) loadings. Finite element mesh is shown in (B).

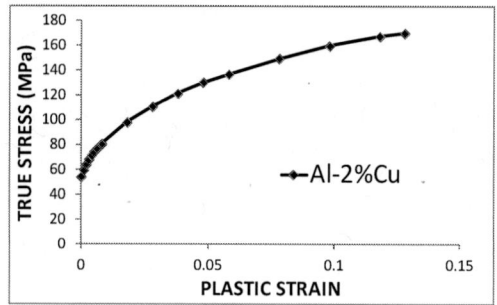

**Figure 3.**   Plastic stress-strain constitutive behavior used as model input for Al-2% Cu matrix.

Cohesive zones were used to model the behavior and failure of the interface between the $Al_2O_3$ fibers and the Al-2%Cu matrix. The constitutive behavior of the cohesive elements was modeled using a traction-separation law, which is shown schematically in Fig. 4. This model assumes linear elastic behavior up to the point at which the interface strength is reached and at which damage is initiated. Initiation of damage, a scalar damage evolution variable ($D$) is introduced which describes the rate at which the material stiffness is degraded. In this work a maximum nominal stress criteria was considered for damage initiation. This criterion is expressed as:

$$\max\left\{\frac{\langle t_n\rangle}{t_n^o}, \frac{t_s}{t_s^o}, \frac{t_t}{t_t^o}\right\} \quad (1)$$

where $t_n$, $t_s$, and $t_t$ are the pure normal, and two shear components of stress at the interface, and $t_n^o, t_s^o, t_t^o$, are the maximum allowable values of the nominal stress. The Macaulay brackets $\langle\ \rangle$ indicate that a purely compressive stress state does not initiate damage.

The damage evolution variable of the cohesive zone elements evolves monotonically from 0 to 1. In this work a linear degradation in stiffness is assumed in which the normal and shear components of the traction-separation material behavior are evolved by the following relations:

$$t_n = \begin{cases}(1-D)\bar{t}_n, & \bar{t}_n \geq 0 \\ \bar{t}_n, & otherwise\end{cases}, \quad (2)$$

$$t_s = (1-D)\bar{t}_s,$$

$$t_t = (1-D)\bar{t}_t$$

where $\bar{t}_n$, $\bar{t}_s$, and $\bar{t}_t$ are the stress components of the undamaged material and damage is assumed to only progress when normal tractions are tensile $(\bar{t}_n \geq 0)$. Damage evolution is defined in the model based on the energy dissipated as a result of the damage process and is equal to the area under the traction-separation curve $(G_c)$. These parameters can be extremely difficult to measure directly from experiments and in this work the cohesive zone parameters, $t_n$ and $G_c$ were inversely determined by calibrating the simulation results to fit the experimentally obtained stress-strain curve for the transverse uniaxial tension test in which failure is dominated almost exclusively by fiber-matrix interface separation which is shown in (Fig. 5A).

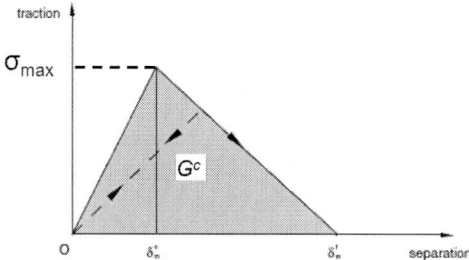

**Figure 4.**  Schematic of traction-separation law used to model progressive damage and failure of matrix-particle interfaces.

## 3. Results and Discussion

The cohesive zone parameters for the interfacial strength were determined by iteratively simulating the transverse uniaxial tension test to fit the result to the experimental curve. Based on this process the final stress-strain curve obtained using cohesive parameters of: $t_n^o$ =155 MPa, $t_s^o = t_t^o$ =89.5 MPa, and $G_c$=75 J/m$^2$ is presented in Fig. 6 along with the experimental result. The result from a simulation without cohesive interface failure is also included for reference which indicates an upper bound on the strength and demonstrates the significant reduction in overall strength when the interfacial failure mechanism is included in the simulations. There is a deviation in the initial yielding portion of the stress-strain curves with the simulation over predicting the experiment slightly. This can at least be partially explained by processing induced defects in the material such as micro porosity and inclusions that are not accounted for in the present model.

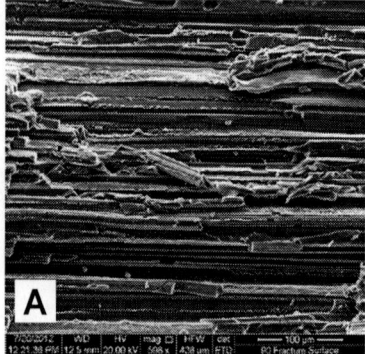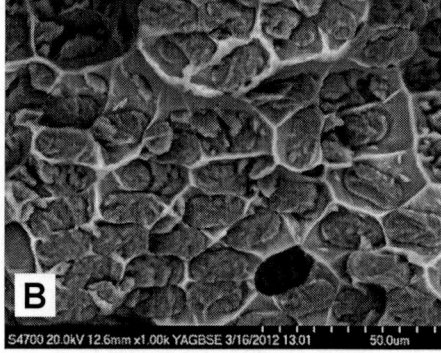

Figure 5.  Fracture surfaces of transverse (A) and axial (B) tension test specimens of unidirectional continuous Al$_2$O$_3$ fiber reinforced Al-2%Cu.

The cohesive parameters obtained from calibration of the transverse tension test were used to simulate the response of the material in the fiber (axial) direction. These simulation results are presented in Figure 7 along with results for a simulation considering a perfectly bonded interface (no cohesive interface failure). The transverse and axial results and presented together on the same scale in Figure 8 to clearly illustrate the dramatic difference in properties in the two loading directions. In the fiber direction the MMC exhibits a bi-linear behavior up to failure. After a very small initial elastic period the slope changes as the matrix begins to plastically deform but the fibers still bear most of the load. Although failure in the axial direction is dominated by fiber failure (Figure 5B) it can be observed that interface failure does contribute significantly to the overall composite response. Compared to the perfectly bonded simulation, the response is softened by interface failure and results in a simulation result that is more closely aligned with the experimental results. These results demonstrate that using the parameters calibrated from the transverse tension experiment provide a reasonably accurate prediction of the composite response when the transverse failure is dominated by interfacial separation. These results provide a first order validation of the calculated cohesive parameters but additional experimentation and simulations need to be considered to completely characterize and validate the constitutive response of the metal-fiber interface. It is noted that there was significant variation in the experimental strain to failure of the axial test specimens which is represented by the dashed line in Figure 7 with the predicted failure strain lying at the high end of the range. This is expected based on previously published data on unidirectional reinforced MMC [7] due to the brittle nature of the $Al_2O_3$ fibers and their inherent sensitivity to internal flaws which follow a Weibull type distribution [3,4] and the strength of the composite will be limited by the lower strength fibers which will fracture first [8]. It is important to note that the current simulations use a fixed average strength which will only predict a single failure strain which is not realistic for these materials which will always have some quantifiable distribution in properties. In order to predict statistically representative MMC properties in will be necessary to include the variation in fiber properties in the constitutive models which is planned for future work.

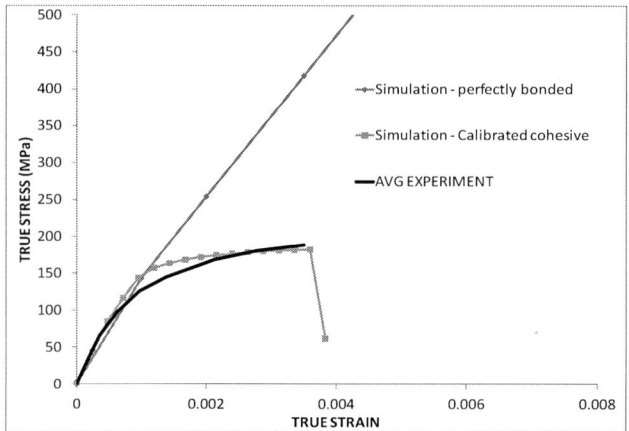

Figure 6. Comparison of experimental results with simulation predictions for perfectly bonded (no interface failure) and simulation predictions with calibrated interface strength for transverse tension test.

To explore the effect of fiber strength on the composite properties and quantify what the maximum performance that could be obtained by this material system by using only high strength fibers, an axial simulation was conducted assuming a fiber strength of 4 GPa which corresponds to fiber strengths in the upper range of experimentally measured values [3,4,6]. The simulation results match exactly the previous but extend the curve and show that a 60% increase strength and a 62.5% increase in strain to failure of the MMC could be obtained by increasing the average fiber strength by 35%.

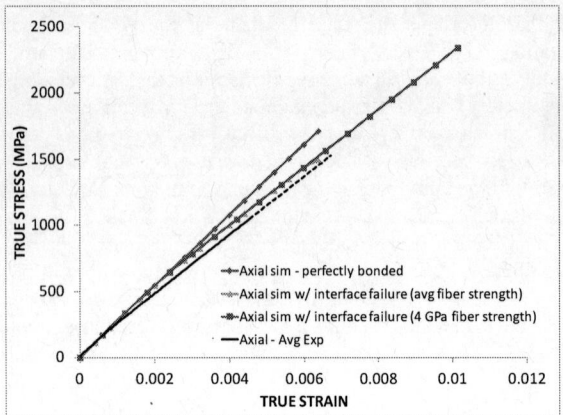

**Figure 7.** Comparison of experimental results with simulation predictions for perfectly bonded (no interface failure) and simulation predictions with calibrated interface strength for axial tension test.

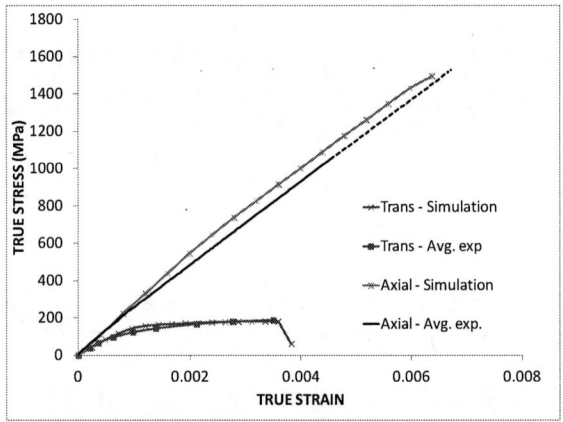

**Figure 8.** Comparison of experimental and numerical generated stress-strain curves for unidirectional continuous $Al_2O_3$ fiber reinforced Al-2%Cu in transverse and axial loading directions.

## 4. Conclusions

A finite element micromechanical model was used to study the deformation response of a unidirectional continuous ceramic fiber aluminum matrix composite up to and including failure behavior. The model incorporates the three dominant failure mechanisms present in MMC systems: ductile matrix failure, cohesive interfacial separation, and brittle fiber fracture. A set of validated constitutive parameters for the cohesive interface were successfully determined from a transverse tension test which was experimentally characterized to fracture predominantly by failure of the fiber-matrix interfaces. The calibrated parameters were validated by comparing simulation and experiments of a tension test in the fiber (axial) direction. The directional dependent plasticity constitutive behavior determined from this modeling effort can be transitioned to a mesoscale laminate or fabric model to predict the continuum response of metal matrix composites.

### Acknowledgements

Financial support for this work provided by an Oak Ridge Institute for Science and Education (ORISE) post-doctoral fellowship at the U.S. Army Research Laboratory.

### References

1. N. Chawla, K.K. Chawla, Metal Matrix Composites, Springer Science + Business Media, Inc., NY, USA, 2006.

2. Mortensen A, Llorca, *Ann. Rev Mater Res* 40 (2010) 243-270.

3. Dibelka J. A. and S. W. Case, "Characterization of Continuous Nextel™ 610 Fiber Reinforced Aluminum," presented at SAMPE, Salt Lake City, UT, October 2010.

4. Wilson, D, Statistical tensile strength of Nextel 610 and Nextel 720 fibres. *Journal Materials Science* 32 (1997) 2535-2542.

5. 3M Nextel™ Ceramic Textiles Technical Notebook, Ref. No. 98-0400-5870-7, 2004.

6. Wilson, D.M. and L.R. Visser, "High performance oxide fibers for metal and ceramic composites," *Composites: Part A* 32 (2001) 1143-53.

7. Ramamurty, U., et al, "Strength variability in alumina fiber-reinforced aluminum matrix composites," *Acta Metallurgica* 45(11), 4603-4613, 1997.

8. Dibelka, J. A. and S. W. Case, "Damage evolution model for hybrid metal matrix composites," presented at TMS 2012, Orlando, FL, March 2012.

# DEVELOPMENT OF PERCUSSION DIAGNOSTICS IN EVALUATING 'KISS' BONDS BETWEEN COMPOSITE LAMINATES

Scott L. Poveromo[1], James C. Earthman[1]

*Department of Chemical Engineering and Materials Science*
*University of California, Irvine*
Irvine, CA 92697, USA

### Abstract

Conventional nondestructive (NDT) techniques used to detect defects in composites are not able to determine intact bond integrity within a composite structure and are costly to use on large and complex shaped surfaces. To overcome current NDT limitations, a new technology was developed based on percussion diagnostics to better quantify bond quality in fiber reinforced composite materials. Results indicate that this technology is capable of detecting weak ('kiss') bonds between flat composite laminates. Specifically, the local value of the loss coefficient determined from quantitative percussion testing was found to be significantly greater for a laminate that contained a 'kiss' bond compared to that for a well bonded sample. Laminates bonded together using supported epoxy film adhesives will be characterized.

<u>Keywords</u>:  composites, kiss bonds, nondestructive testing, percussion

## Introduction

Bonding composite structures together using adhesives provides many advantages over other joining methods. These advantages include distributing the load over a large bond area, reduced weight, and ability to join dissimilar materials together, higher stiffness and toughness over the bond area and in many cases lower manufacturing cost [1, 2]. However, one of the limitations when using adhesives is the inability to determine non-destructively if the bond joint assembled meets structural requirements. Unfortunately, this leads to a conservative design approach and applying fasteners through the bond to ensure joint integrity. The addition of fasteners increases cost and weight and greatly diminishes the aforementioned advantages. Furthermore, fasteners cannot be the solution to attach materials that are being bonded within composite laminates that act as toughening aids or sensors for creating multifunctional materials. In order to address this limitation, a new NDT method needs to be developed that can detect adhesive 'kiss' bonds, where the adhesive shear strength is low due to contamination on the bonding surfaces or improper handling, mixing, or curing of the adhesive. It is believed that most 'kiss' bonds result from poor surface preparation of as-molded surfaces due to excess fluorocarbons, silicones, plasticizers, etc. introduced from the manufacturing process [2,4]. If these contaminants are not removed, they will decrease the surface energy which will decrease the contact angle between the adhesive and the bonding surface causing a decrease in shear strength. As a result, the bond that is formed will not be able to carry load as both substrates will in essence be 'kissing' one another.

The common nondestructive testing (NDT) techniques used to detect defects in composite structures are ultrasound, thermography, X-radiography and shearography [4-14]. However, these techniques are not able to detect kiss bonds within a composite structure and are either costly or difficult to use on large and complex shaped surfaces. A recent method developed by Bossi et al. uses a high peak powered pulsed laser to provide a localized dynamic proof test for bonded structures [15]. This laser bond method has shown to be able to detect differences in adhesive shear strength; it is destructive if the adhesive bond is weak and is non-destructive if the bond is strong. However, this method cannot be used in hard to access areas and is expensive to set-up. One cost effective method that has shown promise to detect differences in bond integrity of aluminum and steel lap shear joints is vibration analysis [16-19]. In vibration analysis, lap shear specimens are vibrated in a fixed free condition where one end of the specimen is fixed and the other end is free to vibrate. An impact force was applied to the specimen and the frequency response was measured. From the frequency response, the loss modulus can be computed dividing the resonant frequency by the half power band width. Changes in loss modulus or resonant frequency can then be compared to the shear strength of different metal to metal bond joints.

Srivatsan et al. examined the effects of percent debond (released) area on the strength of steel double lap shear specimens bonded together with epoxy adhesive. The authors discovered as the percent debond area increases, both the loss modulus and resonant frequency showed significant changes. As the percent debond area increased, the loss modulus increased and the resonant frequency decreased, while the joint strength decreased. The authors attributed the increases in damping capacity to friction caused by the opening and closing of the debonded interfaces during the vibration test. The decrease in resonant frequency is attributed to the decrease in stiffness of the overall bonded structure due to the higher debond area. Also, Yang et al. created shear strength reductions in structural reaction injection molded (SRIM) double lap shear joints bonded together with polyurethane adhesive by varying the percent bond area that was sanded [2]. The authors also found as the percent bond area that was sanded decreased, the loss modulus increased and the shear strength decreased.

Although the results from the vibration analysis technique are promising, the measurements were taken on homogenous adherends and subsequently easier to examine than laminated fiber reinforced composite substrates. Also, vibration analysis must be performed on specific specimen geometry and is limited to a small specimen size.

A cost effective legacy NDT method that has been used extensively in the past to inspect large composite components is the so-called coin tap test [20]. The coin tap test has been used in the composites industry to differentiate between a well bonded, defect-free composite part to one that has defects [20]. This method uses a coin or the back of a hammer to impact the composite part manually (a light tap). A sound of a 'good' section is very crisp and clear and radiates, while a sound from a 'bad' area has a dull thud. Distinguishing between sounds can be difficult even for a trained technician and the use of this technique over long periods of time is impractical. Also, no property information is gathered using this method that will help determine the extent of damage of the composite being inspected.

Besides using sound, a tap test can be used to identify defective areas in a composite by examining different parameters controlling the impact response. This paper describes a percussion technique developed to detect the difference between a well bonded composite structure (high shear strength) and a poorly bonded composite structure (low shear strength). Also, the technique is not dependent on specific specimen size and is not costly to operate. Specifically, the loss modulus, impact force and transmitted force will be measured and compared for both well bonded and poorly bonded composite specimens. The objective of this paper is to show the potential of using percussion diagnostics as a NDT method to inspect adhesive bond joints between composite substrates.

## Quantitative Percussion Testing

Quantitative percussion instrumentation (Periometer®, Perimetrics, LLC) was evaluated in the present work for determining bond integrity in composite laminate samples. This system was originally developed to investigate and compare the damping capacity of human teeth and dental implants [21-24]. As shown in Figure 1, a stainless steel rod containing a force sensor is accelerated by an electromagnetic coil to a pre-determined velocity just prior to impact.

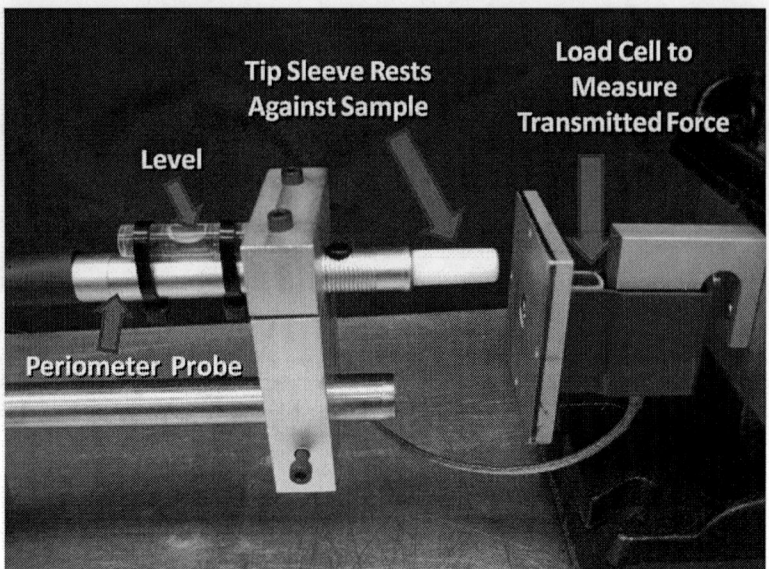

**Figure 1.** Photograph of percussion probe (Periometer) and load cell used in the present work.

While the percussion rod is in contact with the specimen, the electromagnetic coil is inactive so that only the kinetic energy of the rod is administered to the specimen. The low level impact of the percussion rod generates a nondestructive stress wave that propagates through the specimen. Inelastic deformation or damping of strain energy during impact is characterized by the loss coefficient, $\eta$, a common damping capacity parameter. Custom software on a computer interfaced to the percussion rod determines the mechanical energy returned to the rod from the force measured vs time for 10 of the 16 impacts. The energy dissipated by the specimen, $D$, is determined from

$$D = U - E_R - D_p \qquad [1]$$

where $U$ is the total strain energy, $E_R$ is the mechanical energy returned to the sensor in the percussion rod and $D_p$ is energy dissipated by sources external to the specimen [23]. The loss coefficient is given by

$$\eta = 1/2\pi * (D/U) \qquad [2]$$

where the total strain energy, $U$, for the present percussion testing is approximately equal to the total kinetic energy prior to impact from the rod ($1/2\ mv^2$). In addition to displaying the loss coefficient, the output provided by the custom software is a plot of the force measured during each impact event (probe force) as well as the transmitted force measured by the force transducer on the opposing surface of the specimen plotted vs time. The average maximum values of probe force and transmitted force are also displayed.

Data compiled using the Periometer have been found to be useful when analyzing damaged composite laminates. For example, Stanley et al. [21] previously used this device to examine damage of a carbon fiber/epoxy laminate induced during standard drop weight impact tests. They discovered that the loss coefficient increased as the drop weight impact energy increased and as the measurement distance to the drop weight impact site decreased (see Figure 2). The reasoning behind this trend is that the damage induced by a drop weight test consists of considerably more sources of internal friction caused by the movement of crack and delamination surfaces. Also, the breaking of fibers results in more stress being supported by the matrix that generally gives rise to greater inelastic deformation and thus more energy dissipated. Fiber sliding also can occur, which increases internal friction and the amount of energy dissipated as heat. Hence, a relationship between internal delamination/micro-cracking and local changes in damping capacity can be exploited to investigate internal damage in composites.

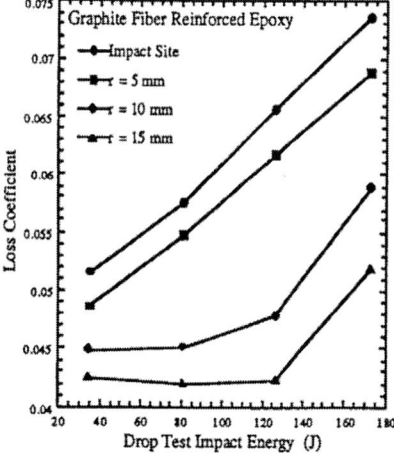

**Figure 2.** Loss coefficient vs drop test impact energy for fiber reinforced composite. From B.D. Stanley, L. Bustemante, and J.C. Earthman, "Novel Instrumentation for Rapid Assessment of Internal Damage in Composite Materials," in Nondestructive Evaluation and Material Properties III (1996), P.K. Liaw, O. Buck, R.J. Arsenault, and R.E. Green, Jr., eds., p. 99, Figure 3. Copyright (c) 1996 by The Minerals, Metals & Materials Society. Reprinted with permission.

### Experimental Procedures

In order to assess the validity of using percussion to detect 'kiss' bonds in composite structures two composite laminates were fabricated; one bonded following a poor surface preparation technique, while the other was bonded per common standard practices. First, two pre-cured carbon fiber/epoxy matrix laminates 305 mm x 305 mm x 1.59 mm (12 in. x 12 in. x 0.0625 in.) were bonded together with a 121 °C (250 °F) cure supported epoxy film adhesive. One specimen had release agent applied in a 152 mm x 152 mm (6 in. x 6 in.) area in the center of both laminates, which simulated a 'kiss' bond while the other bonded laminate had no release agent applied. To ensure the release agent created a poor bond, it was baked onto the laminate surface prior to bonding.

Before testing, the Periometer® was powered on and left idle for 15-30 minutes to eliminate the possibility of electronic drift effects during testing. Next, the Periometer was calibrated using two different materials with known loss coefficient values; aluminum alloy 6061-T6 rod and polytetrafluoroethylene (PTFE). The loss coefficient was measured three separate times on each of these materials and the data were used to calculate the effective elastic modulus of the rod and the amount of energy dissipated by external sources (not the specimen) such as the probe [21]. Once these values were computed, the system was calibrated and testing could begin. Percussion was performed on locations in the center portion of the panel and the impact/time curves for both specimens were recorded. The loss modulus and probe (impact) force were measured using the fixture shown in Figure 3.

**Figure 3.** Photographs of fixture holding the 305 mm x 305 mm (12 in. x 12 in.) bonded laminates tested using the present (a) percussion probe and (b) load cell. The specimen was preloaded by pressing it against the backing plate on the load cell by the probe and positioning bolts shown.

The transmitted force was measured using a 44.5 N (10 lbf) load cell on the side of the specimen opposite to where the percussion was performed. The load cell was attached to a 50.8 mm x 50.8 mm (2 in. x 2 in.) backing plate. The backing plate maintained uniform contact with the specimen and facilitated the measurement of a representative transmitted force by the load cell. An illustration of where the backing plate was located relative to the impact locations is illustrated in Figure 4. Prior to transmitted force measurements, the laminates were machined to help center the load cell on the backside as shown in Figure 4.

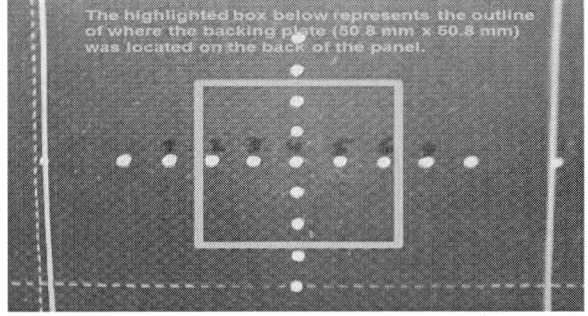

**Figure 4.** Illustration of location of load cell backing plate (orange outline) relative to the impact locations.

After percussion measurements were taken, the two panels were examined using ultrasound in pulse-echo mode at Northrop Grumman. A C-scan of each panel was taken at 5 MHz using a 2" focused transducer as shown in Figure 5.

**Figure 5.** Illustration of ultrasonic tank set-up in pulse-echo mode.

## Percussion Results

### Loss Coefficient and Impact Force Measurements

Impact measurements were taken at seven different locations horizontally across the middle of the two composite laminates as shown in Figure 6. A summary of the results for this testing is illustrated in Figures 6 and 7. The loss coefficient was found to be higher at every location tested for the released ('kiss' bond) panel when compared to the not-released (well bonded) panel. The maximum probe force (shown in Figure 7) is also lower for all locations located on the released panel. A single factor analysis of variance (ANOVA) was completed for the loss coefficient and maximum probe force between data collected for released versus the well bonded specimens. The p values obtained for loss coefficient and probe force data were between 6 to 8 x $10^{-5}$ indicating significant difference between the results for the released and not-released panels. The significantly greater loss coefficient and corresponding lower decrease in probe force can be explained by greater internal friction within the poorly bonded specimens. This explanation is consistent with damping capacity data generated for metal adherends [17-18].

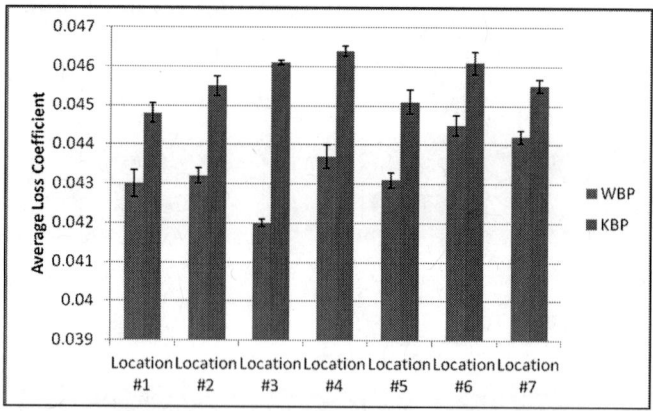

**Figure 6.** Loss coefficient measured at different locations for the well bonded-not-released panel (WBP) and the released 'kiss' bond panel (KBP).

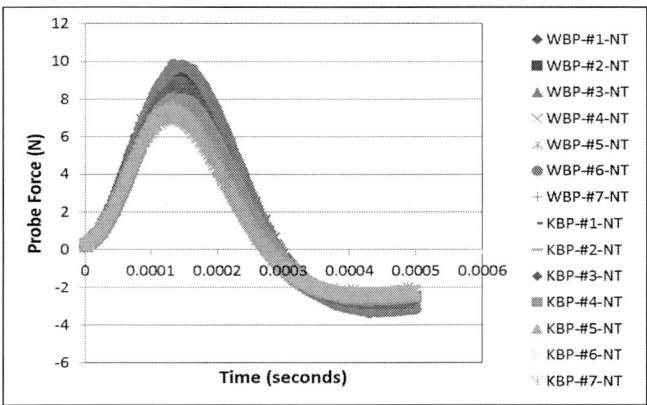

**Figure 7.** Probe force vs time measured at 7 different locations for the well bonded not-released panel (WBP) and the released 'kiss' bond panel (KBP).

## Transmitted Force Measurements

For transmitted force measurements, a pre-load was needed to ensure good intimate contact between the backing plate and sample. However, the pre-load was found to have an effect on the transmitted force measurement as well as the loss coefficient and probe force measurements. A low pre-load of 3 to 4 N was used to create contact between the backing plate and sample but not too high to change the alignment of the sample during testing. The results of the transmitted force measurements are shown in Figure 8. Most of the maximum transmitted force measurements were higher on the not released (well bonded) panel compared to the released ('kiss' bonded) panel. This was especially true for the three central locations where the well bonded panel recorded significantly higher transmitted force values. The explanation for this trend could be that the impact energy will have a harder time propagating through the thickness of two laminates that are poorly bonded at the interface. The well bonded laminates should be able to transmit force easier from the front to back surface. The p-value calculated in a completed analysis of variance for the three central locations was 0.0013.

The maximum transmitted force also varied as a function of percussion location on the specimen as shown in Figure 7. The transmitted force was higher at locations #1 and #2 for the KBP when compared to the middle locations (#3, #4, #5). This trend was not observed for the WBP. It appears that locations 1, 2, and 7 were sufficiently close to the edge of the kiss bond so that greater forces can still be transmitted to the load cell.

The probe force and loss coefficient were also measured with the backing plate in contact with the WBP and KBP samples at a pre-load of 3 to 4 N. Table 1 shown below summarizes the average maximum probe force and average loss coefficient measured with corresponding p-values. The average probe force and loss coefficient were similar in value for WBP and KBP samples.

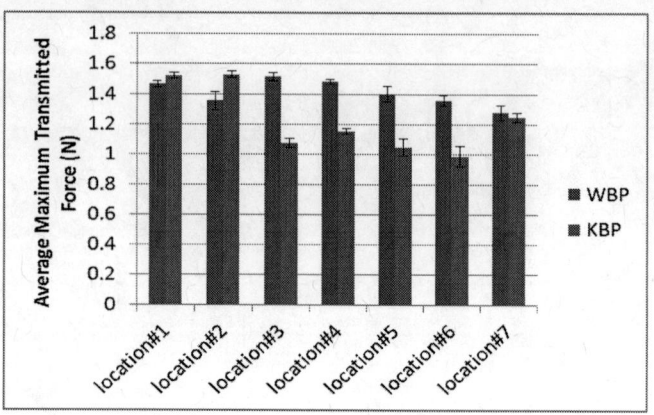

**Figure 8.** Average maximum transmitted force measured at 7 different locations using a low pre-load (3 to 4 N) for the well bonded-not-released panel (WBP) and the released 'kiss' bond panel (KBP). A significant difference (p = 0.0013) in transmitted force was determined for the three central locations (3 through 5).

**Table 1.** Average loss coefficient and probe force measurements for WBP and KBP tested at low pre-load conditions. Average values reported are the overall average of the measurements taken at each of the seven locations across the sample.

| Specimen | Pre-load | Avg. Loss Coefficient | Single Factor ANOVA P-value | Avg. Max Probe Force (N) | Single Factor ANOVA P-value |
|---|---|---|---|---|---|
| WBP | Low | 0.0338 | 0.667 | 9.07 | 0.780 |
| KBP | Low | 0.0351 | | 9.16 | |

**Comparison of Transmitted Force and Impact Force Measurements**

Transmitted force and probe force data for both a WBP and a KBP at location #4 are shown in Figure 9. No difference was observed in the probe force plots. However a difference in the behavior of the transmitted force curves was observed after 0.001 seconds. The transmitted force for the WBP was slightly higher than for the KBP after 0.001 seconds until both force curves attenuated at about 0.003 seconds. A difference in transmitted force between these panels was expected as the WBP should record a higher transmitted force.

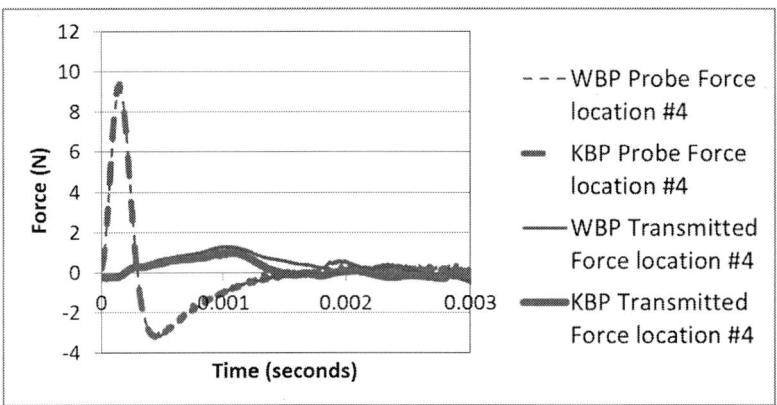

**Figure 9.** Probe and transmitted force vs time measured at location #4 using a low pre-load (3 to 4 N) for the well bonded panel (WBP) and the released 'kiss' bond panel (KBP).

**Ultrasonic Testing Results**

The results from the ultrasonic testing performed on both well bonded and released 'kiss' bond panels are summarized in Figures 10 and 11. Three gates were set-up when analyzing the data to help characterize the bonded panels. Gate 1 represents the front of the bondline, Gate 2 represents the back of the bondline and Gate 3 represents an image of the internal skin. Very little change in % amplitude was observed for the well bonded panel in all three Gate areas. A typical A-scan showed a very small change in amplitude within the bondline area (between gates 1 and 2). Also, the bondline thickness was measured by subtracting time of flight (TOF) measurements between Gates 1 and 2. A histogram of the TOF measurements showed the film thickness to be centered around 0.009". The released 'kiss' bond panels illustrated an observable change in percent amplitude over the released area. The images of the front and back bondline show significant light areas in the center of the panel signifying the panel had disbonded. Typical A-scans show a large change in the wave amplitude over the disbonded area compared to the well bonded area.

The disbond observed in Figure 11 may have resulted from the trimming operation performed prior to transmittance force measurements. The panels were trimmed using a cut-off saw which would create an air gap around the released area which ultrasound would detect. Another reason why ultrasound was able to pick-up the poorly bonded area is due to the differences in thermal expansion (CTE) between the epoxy film adhesive and cured carbon fiber/epoxy laminates in the lateral directions. When the laminate is heated to 250F during the cure of the adhesive, the film adhesive will expand while the laminates will remain relatively stationary. Upon cool down the change in strain of the film adhesive could cause the adhesive to disbond from the released area on the laminate.

To determine if this is repeatable another two composite specimens were fabricated following similar manufacturing procedures. Two pre-cured carbon fiber/epoxy matrix laminates 266.7

mm x 190.5 mm x 1.59 mm (10.5 in. x 7.5 in. x 0.0625 in.) were bonded together with a 121 °C (250 °F) cure supported epoxy film adhesive. One specimen had 5 coats of release agent applied in a 114.3 mm x 190.5 mm (4.5 in. x 7.5 in.) area in the center of both laminates, which simulated a 'kiss' bond while the other bonded specimen had no release agent applied. The results from the ultrasonic testing performed for the second set of panels are shown in Figures 12, 13 and 14. The same three gates were set-up when analyzing the data to help characterize the bonded panels. Gate 1 represents the front of the bondline, Gate 2 represents the back of the bondline and Gate 3 represents an image of the internal skin. The second well-bonded panel illustrated similar results to the first well bonded panel tested as there was very little change in % amplitude across the center of the panel. A significant change in percent amplitude over the released area on the kiss bond panel was observed. The large red section of the released area illustrates a disbond while the red/yellow speckled areas represent intermittent weak bonded areas. The red/yellow speckled pattern observed may be due to the surface roughness of the peel ply surface that the release agent was applied to. The release agent may not have covered the bonding surface completely causing the speckled pattern C-scan image. Typical A-scans show a large change in the wave amplitude over the red disbonded area (Figure 14) compared to the blue well bonded area (Figure 13).

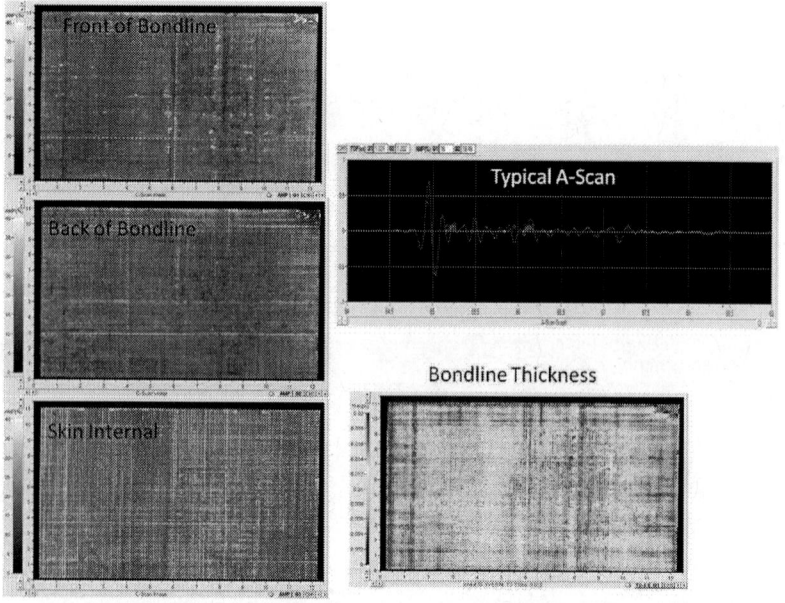

**Figure 10.**  C-scans and A-scan of the well bonded panel.

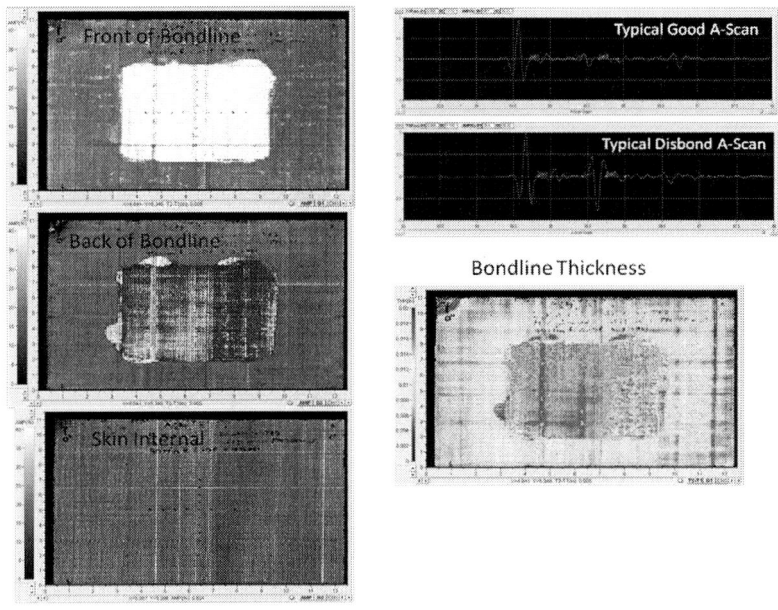

**Figure 11.** C- scans and A-scans of the 'kiss' bonded panel.

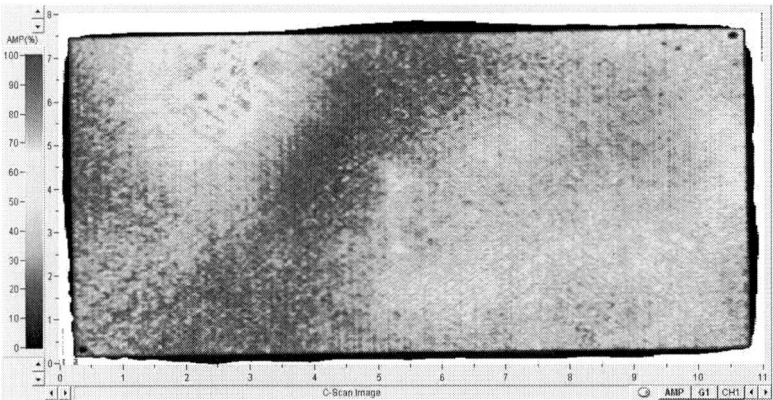

**Figure 12.** C-scan of second well-bonded specimen.

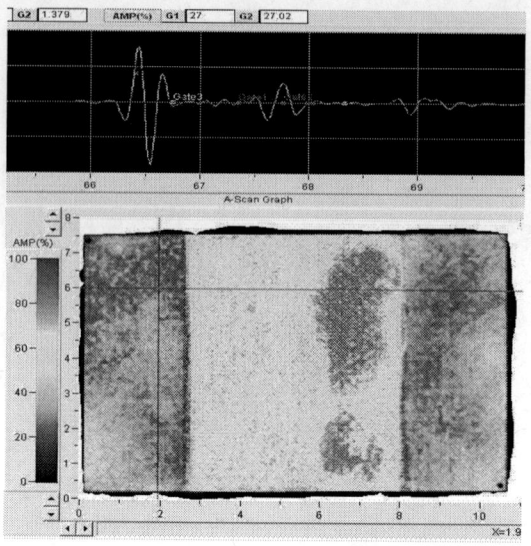

**Figure 13.** C- scan of second 'kiss' bonded panel and typical good A-scan (blue area).

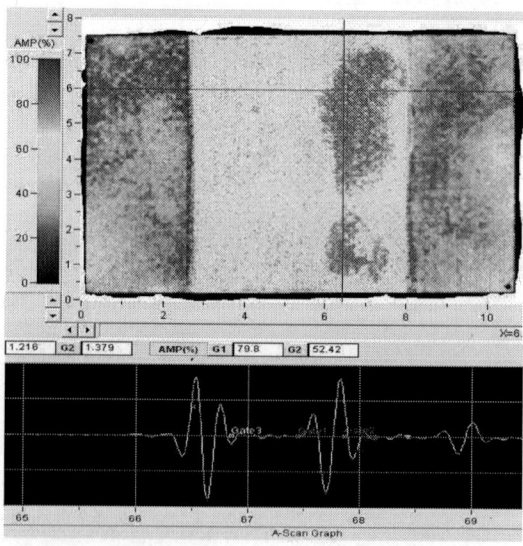

**Figure 14.** C-scan of second 'kiss' bonded panel and typical poor A-scan (red area).

# Conclusions

## Concluding Remarks

A significantly greater loss coefficient determined from percussion testing was observed for a 'kiss' bonded panel compared to that for a well bonded panel. Also, a significantly lower maximum probe force was observed for the 'kiss' bonded panel compared to that for a well bonded panel. The increase in loss coefficient and decrease in probe force are thought to be due to greater internal friction during the percussion for poorly bonded specimens when not using the load cell backing plate on the backside of the test specimen.

Using a low pre-load force (load cell backing plate in contact with the test specimen), a significantly greater maximum transmitted force was observed for the well bonded panel compared to that for a 'kiss' bonded panel. The maximum transmitted force was higher for locations closer to the edge of the kiss bond when compared to locations near the center of the kiss bond region. The greater transmitted force for the well bonded panel under a low pre-load appears to be the result of the two bonded substrates having a greater ability to transmit load across the interface when compared to the 'kiss' bonded panel.

Disbonds between the film adhesive and laminates in the released areas were observed in C-scan images from ultrasonic testing in pulse-echo mode at 5 MHz. The disbond areas occur due to differences in CTE between the film adhesive and carbon fiber/epoxy laminates upon cool down from the adhesive cure temperature. Machining and/or excessive handling of laminates with kiss bonds may also cause the disbonds to form. Follow up mechanical testing shall be conducted to verify the shear strength of the adhesive for all four bonded specimens inspected.

## Acknowledgements

The authors would like to thank Northrop Grumman for providing the four bonded laminates tested and Axiom Materials for providing laminates used in the testing and consultation throughout the project. The authors would also like to thank Bill Sheppard for his help in performing the ultrasonic testing. Financial support from Northrop Grumman for ongoing and future work is also gratefully acknowledged.

## References

1. Barnes TA, Pashby IR. "Joining techniques for aluminium spaceframes used in automobiles Part II- adhesive bonding and mechanical fasteners." *J Mater Process Technol* 99 (2000): 72-79.

2. Yang S, Lan G, and Gibson RF. "Nondestructive detection of weak joints in adhesively bonded composite structures." *Compos Struct* 51 (2001): 63-71.

3. Wingfield JRJ. "Treatment of composite surfaces for adhesive bonding." *Int J. Adhes Adhes* 13 (1993): 151-156.

4. Nagy PB. "Ultrasonic detection of kissing bonds at adhesive interfaces." *J. Adhes Sci. Technol* 5 (1991): 619-630.

5. Dragan K and Swiderski W. "Studying Efficiency of NDE Techniques Applied to Composite Materials in Aerospace Applications." *Acta Phys Pol A* 117 (2010): 878-883.

6. Hsu, D.K. "Nondestructive Inspection of Composite Structures: Methods and Practice." *17th World Conference Nondestructive Testing*. Shanghai, China, October 25th-28th, 2008.

7. Ray BC. et al. "Evaluation of Defects in FRP Composites by NDT Techniques." *J Reinf Plast Comp* 26 (2007): 1187-1192.

8. Haczmarek H. "Ultrasonic Detection of Damage in CFRPs." *J Compos Mater* 29 (1995): 59-95.

9. Gros XE. "Current and future trends in non-destructive testing of composite materials." *Ann Chim Sci Mat* 25 (2000): 539-544.

10. Scott IG and Scala CM. "A review of non-destructive testing of composite materials." *NDT Int* (1982): 75-86.

11. Georgeson G. "Recent Advances in Aerospace Composite NDE." *Proc of SPIE* 4704 (2002): 104-115.

12. Mayr G. et al. "Active thermography as a quantitative method for non-destructive evaluation of porous carbon fiber reinforced polymers." *NDT&E International* 44 (201): 537-543.

13. Avdelidis NP. "Infrared thermography as a non-destructive tool for materials characterisation and assessment." *Proc. of SPIE* 8013 (2011): 801313-1-801313-7.

14. Moore D, et al. "2009 DOE Vehicle Technologies Program Review." http://www1.eere.energy.gov/vehiclesandfuels/pdfs/merit_review_2009/lightweight_materials/lm_15_moore.pdf. Date site accessed 07/20/12.

15. Richard Bossi, et al. "Laser Bond Testing." *Materials Evaluation* (2009): 819-827.

16. Pandurangan P and Buckner GD. "Non-destructive evaluation of metal-to-metal adhesive joints using vibration analysis: experimental results." *Proc. of SPIE* 6176 (2006): 61760G-1-61760G-9.

17. Shiloh K et al. "Aging Characterization of adhesives and bonded joints by non-destructive damping measurements." *J Intel Mat Syst. Str* 10 (1999): 353-362.

18. Srivatsan TS et al. "Electromagnetic measurement of damping capacity to detect damage in adhesively bonded material." *Materials Evaluation* 47 (1989): 564-569.

19. Chung DDL. "Structural composite materials tailored for damping." *J Alloy Compd* 355 (2003): 216-223.

20. Cawley P and Adams RD. "The mechanics of the coin-tap method of non-destructive testing." *Journal of Sound And Vibration* 122 (1988): 299-316.

21. Stanley BD, Bustemante L and Earthman JC. "Novel instrumentation for rapid assessment of internal damage in composite materials." *Nondestructive Evaluation and Materials Properties III* (1997): 97-100.

22. Brenner DA and Earthman JC. "Novel instrumentation for quantitative determination of energy damping in materials and structures." *Scripta Metallurgica et Materialia* 31 (1994): 467-471.

23. Lincoln J, Rieger LE and Earthman JC. "Instrumentation for determining the local damping capacity in honeycomb sandwich composites." *Journal of Testing and Evaluation* 34 (2006): 232-236.

24. United States Patent # 6120466. "System and method for quantitative measurements of energy damping capacity."

# ENHANCED MECHANICAL PERFORMANCE OF WOVEN COMPOSITE LAMINATES USING PLASMA TREATED POLYMERIC FABRICS

Timothy R. Walter[1], Andres A. Bujanda[1], Victor Rodriguez-Santiago[1], Jacqueline H. Yim[1], Jose A. Baeza[2] and Daphne D. Pappas[3]

[1] The United States Army Research Laboratory (ARL)
4600 Deer Creek Loop, APG
MD 21005, USA

[2] Naval Air Station Patuxent River
Patuxent River
MD 20670, USA

[3] EP Technologies LLC
Akron, OH 44311, USA

### Abstract

Interest in composites for structural applications has increased due to the demand for lightweight high performance materials. One factor limiting the use of laminate composites in structural applications is the poor interfacial strength between the fibers and matrix. Due to the propensity of a composite to delaminate under impulse loading, it is necessary to develop composite structures with inherent resistance to such damage. Several methods have been developed to enhance performance including through thickness reinforcements and toughened matrixes. However, these methods do not enhance the fiber and matrix interface strength.

Recently researchers at the U.S. Army Research Laboratory have utilized atmospheric plasma technology to chemically alter the near surface properties of various polymeric fiber systems, enabling stronger, more effective bonding between the fiber and matrix.

In this study a plain woven Kevlar fiber system was treated using a dielectric barrier discharge system and consolidated using a vacuum assisted resin transfer molding process. The composite was evaluated using traditional mechanical test methods to determine response of the composite. The tests include tension, compression and shear testing in the plane of the laminate. The results from both treated and untreated specimens were compared to determine the effect of plasma treatment on the immobilization of the fibers in the polymer matrix during deformation.

**Keywords**: Polymeric Fiber Composites, Digital Image Correlation, Atmospheric Plasma

## Introduction

Aramid fibers such as Kevlar have been used as reinforcement in polymer composite systems since they were introduced commercially in the 1970's. (1) Aramid fibers exhibit excellent high temperature behavior, high elastic modulus, and high strength values which make them excellent reinforcing fibers in a variety of polymer-based composite systems. Additionally, aramid fibers exhibit high chemical resistance, a quality which, when coupled with high crystallinity and lack of polar functional groups in the molecular chain, leads to a fiber surface that is smooth and chemically inert. (2) While desirable qualities for stand-alone applications, the inability of the Kevlar fibers to readily bond to other materials limits its usefulness in composite applications. As a component in a composite system, the ability of the reinforcing Kevlar fiber to form a strong, covalent bond with the matrix material is often the limiting factor in the performance and durability of the system.

Numerous research efforts have been made to modify surface morphology, topography, and chemical reactivity of Kevlar fibers in order to enhance reactivity and interfacial adhesive strength of the fiber/matrix interface. Various methods have been utilized to achieve this characteristic, including acid treatments (3-4), γ-ray and UV/Ozone irradiation (5-7), corona discharges (8), thermal treatments (9-10). Low pressure plasma treatments have also successfully used to modify the aramid surface (11), as well as grafting nano-scale species such as silica nanoparticles (12) and carbon nanotubes for the purpose of increasing frictional energy between fibers, effectively decreasing their mobility. Many of these methods, however, while functionalizing the Kevlar fiber surface, also act to physically degrade the surface and interior portions of the fibers by breakage of the macromolecular chain or by fibril splitting or breaking. This degradation causes a decrease in the mechanical properties of the fiber and resulting composite. (13) It is, therefore, desirable to search for methods to functionalize the surface of the Kevlar fiber without negatively affecting the bulk mechanical properties.

Non-thermal atmospheric–pressure plasmas such as dielectric barrier discharges are becoming increasingly popular as a surface modification tool due to their ability to easily produce stable plasma discharges, being a non-hazardous waste producing operation, low operating costs, fast materials processing speed and industrial scalability (14-15). It has been shown that dielectric barrier discharges (DBD's) have the ability to modify polymeric surfaces via plasma etching, polymerization or cross-linking, by the introduction of chemical functional groups, and by using plasma enhanced chemical vapor deposition processes (PECVD) to deposit functional coatings. Each of these techniques can be tailored to enhance adhesive properties in polymer-based composites. Many varieties of surface functional groups can be achieved depending on the reactive gas used, leading to tailored surface properties favoring enhanced adhesion, biocompatibility, and printability (16-17). Also, the effects of the plasma exposure are only observed on the surface of the Kevlar fiber, causing no change in the bulk mechanical properties of the material. This enables plasma treatment under atmospheric pressure conditions as a viable alternative method to traditional techniques, as it provides a uniform modification, requires short processing times, is not environmentally hazardous, and does not require costly and complex vacuum systems used in low-pressure plasmas (18).

This study showcases the effect of DBD plasma discharges on the surface of scoured Kevlar fabrics and how the changes in surface functionality reflect changes in mechanical behavior of fiber-reinforced polymer matrix composites. He/$O_2$ plasmas (19-20) were used to treat scoured

Kevlar fabric and the treated fabrics were then incorporated into composite panels. Silica nanoparticles were grafted onto one set of plasma treated specimens in order to examine the effect of nanoscale "roughening" on the behavior of the Kevlar composites. The panels were then machined into various mechanical test configurations (tensile, compression, shear) and evaluated for mechanical behavior and load response. It is demonstrated that under the correct plasma conditions, it is possible to achieve increases in strength due to increased interfacial adhesive strength of the fiber/matrix interface and through the grafting of nanoscale particulates on the aramid fiber.

## Experimental Approach

### Plasma Treatments and Materials

The plasma system used in these experiments was an APC 2000 from Sigma Technologies Intl. Inc. The system uses a cylindrical roller configuration and operates in an open atmosphere setup, as shown in Figure 1. The roller serves as the ground electrode and is covered with $Al_2O_3$, the dielectric material. Two high voltage electrodes are positioned on top of the roller at a distance of 2 mm, and have slit channels to allow gas diffusion. The flow rate of the gases was equally divided between electrodes 1 and 2. A radio-frequency power supply operating at 90 kHz was used, with power densities ranging between $0.861–2.58$ W cm$^{-2}$.

Scoured (sizing free) polyaramid (Kevlar) woven fabrics were used in this study. Helium was used as the carrier gas for all plasmas at 12000 sccm. $O_2$ was used as the reactive gas at a flow rate of 60 sccm. These gas mixtures were chosen to achieve the most uniform discharge possible for each plasma condition. Under these conditions the plasma generated was mostly uniform; however, the presence of micro-discharges was visible in some instances.

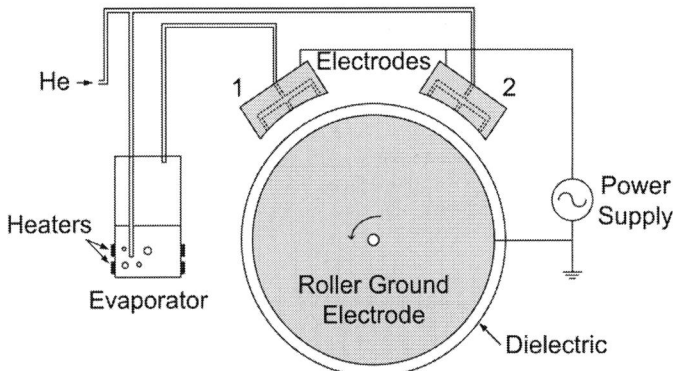

Figure 1.  Schematic of cylindrical roll-to-roll atmospheric plasma system.

### Fiber-Nanoparticle Modification

Plasma treated substrates were functionalized with silica according to the following two-stage procedure. In the first stage, a 1% solution of glycidoxypropyltrimethoxy silane ([GPS], Alfa Aesar, Ward Hill, MA) in methanol was prepared, and the plasma treated films were immersed for one minute. After air drying, the films were placed in an oven to dry and fully cure at 70 °C for 60 min. The second stage of treatment required the preparation of a solution of colloidal silica (Ludox, Aldrich, St. Louis, MO). A 1% solids solution was prepared, using a 90/10 ethanol/water solution with a pH adjusted to 4.5 as the solvent. The GPS-treated films were immersed in the colloidal silica solution for 1 min, and were then allowed to air dry. The films were then dried and fully cured in an oven at 70 °C for 60 min.

**Fiber Reinforced Composites**

Fiber reinforced composites were constructed on a glass tool surface coated with Frekote (Henkel) mold release. The bottom layer of the vacuum assisted resin transfer molding (VARTM) was a Teflon release ply and was used to aid in de-molding of the finished part. The Kevlar fabrics were dry stacked on top of the bottom release ply layer to the desired final thickness. A porous nylon peel ply (Richmond A-8888) was draped on top of the stacked fabrics followed by a layer of distribution media. The distribution media helps to maintain an even distribution of resin on the top of the panel and also facilitates the flow of resin through the thickness of the panel. Sealant tape was placed around the edges of the lay-up. Vacuum grade tubing was used as inlet and outlet vents for resin flow. A bleeder (Airweave by Airtech) was placed at the outlet vent to insure thorough wetting of the dry stacked components of the lay-up. A vacuum bag (Stretchlon 800 by Airtech) was placed over the molding area and sealed firmly to prevent air leaking. Resin infusion was accomplished by the use of a vacuum pump.

The resin used for this investigation was SC15 epoxy manufactured by Applied Polernaric, Inc. (Benica, CA). Once infused into the panels, the epoxy resin was pre-cured and post-cured according to the manufacturer instructions. Finished panels were approximately 12" X 12" X 0.15" thick.

**Machining of Mechanical Test Specimens**

Several specimens were cut from each finished panel. The specimens include tension, compression, and shear, each cut to their respective ASTM recommended geometry using a water-jet cutter. Tension and compression specimens consisted of long narrow strips sectioned along the weft direction. Shear specimens were first sectioned into rectangular coupons in which v-notches were then precision ground to produce the final shape. The specimens were carefully measured to determine the appropriate gage dimensions. The regions of interest were coated with white primer enamel paint and then speckled to produce a pattern used for digital image correlation during testing.

**Mechanical Testing**

Three different tests were performed on each panel to determine the effect of treatment on mechanical response of the composite laminate. The test procedures were adapted from ASTM standards for testing laminate composites and include in-plane tension, compression, and shear testing. These tests are summarized in detail bellow. An Instron (Norwood, MA) 1125 universal test system was used to perform each test. Digital image correlation (DIC) (21-23) was utilized to measure the in-plane strains during each test. The procedures for this technique are also discussed below.

## Tensile

To determine the in-plane tensile properties of the composite laminates tensile test were performed following ASTM D3039 (24). Five specimens measuring 25 mm x 200 mm were sectioned from each plate. Wedge action grips were used to apply tensile force to the specimens. No tabs were used during gripping, however a layer of emery cloth was placed between the specimen and grip faces to reduce the chance of grip induced failure. The specimens were loaded at a constant crosshead rate of 2.0 mm/min. From this test the in-plane tensile modulus ($E_{11}$), offset yield strength ($\sigma_{Tyield}$) and ultimate strength ($\sigma_{Tult}$) were determined. The offset yield strength was determined due to the highly non-linear behavior of the tensile response of the woven composite.

## Compression

Compression tests were performed following ASTM D6641 (25). Coupons measuring 13 mm x 140 mm were tested using a Wyoming Combined Loading Compression fixture, (Wyoming Test Fixtures Inc. Salt Lake City, UT). This method applies a compressive force through a combination of shear on the faces of the specimen and end loading. The center section of the specimen remains unsupported to allow the use of strain gages. However two guide posts obscure the front and back surfaces preventing the use of DIC to determine the in-plane strains during deformation of the specimen. Strain was measured on the side of the specimen, however it was determined that this method did not produce repeatable results and therefore the strain measurements were utilized only for purpose of plotting the trends. From this test the in-plane maximum compressive stress ($\sigma_C$) was determined.

## Shear

Lastly ASTM D5379 (26) was used to determine the shear response of the composites. This method was chosen over other test methods (27) to minimize the material needed to perform the tests. As discussed above coupons were first sectioned from the plate, then, v-notches were machined into each specimen. These specimens were loaded into an Iosipescu Shear Test fixture (23). This fixture supports the top and bottom surfaces to minimize rotation of the specimen. One end of the coupon is fixed while the other is displaced to produce a concentrated shear deformation in the notched region. The fixture was loaded in compression at a constant crosshead rate of 2.0 mm/min. Form this test both the in-plane shear modulus ($G_{12}$) and shear strength ($\tau_{max}$) was determined.

## Digital Image Correlation

DIC is a popular technique for acquiring full field strain history during mechanical testing of many different materials (21-23). In this study stereoscopic DIC was used during the evaluation of the Kevlar-SC15 composites. Two Retiga 2000R, (Q-Imaging, Surrey, BC) cameras were used to produce images during each test. Vic-Snap, (Correlated Solutions, Columbia, SC) was used to record the images. Force and displacement data was output from the Instron controller and also recorded with each image allowing synchronization between DIC strain measurements and stress histories. Correlation was performed using Vic3D (Correlated Solutions, Columbia, SC) which was also used to compute the required strains for each test. A typical correlated result from a shear test is shown below in Figure 2.

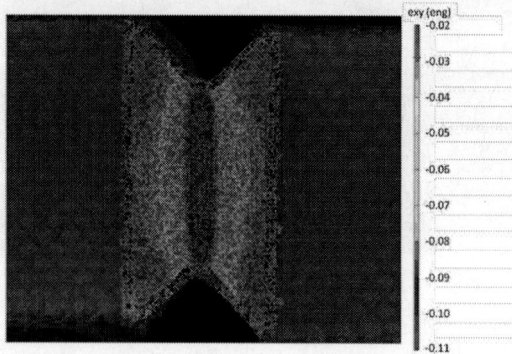

**Figure 2.** Result from DIC showing shear strain in Kevlar composite.

## Results

For each set of tests performed, average values of the resulting mechanical parameters were determined for each panel. These results are shown in Table 1. Overall the results show a distinct affect on the mechanical behavior due to surface modification using atmospheric plasma and colloidal silica. Both modulus and strength were improved with treatments; however, the results were inconsistent in showing improvements during each mode of deformation. A detailed discussion for each test procedure is shown below.

**Table 1.**  Results from mechanical testing composite specimens

| Treatment | $E_{11}$ (GPa) | $\sigma_{Tyield}$ (MPa) | $\sigma_{Tult}$ (MPa) | $\sigma_C$ (MPa) | $G_{12}$ (MPa) | $\tau_{max}$ (MPa) |
|---|---|---|---|---|---|---|
| As-Received | 13.23 | 174.4 | 563.6 | 69.25 | 1.22 | 28.7 |
| Plasma | 17.63 | 192 | 566.6 | 80.33 | 1.03 | 29.27 |
| Plasma+SiO$_2$ | 12.92 | 208.1 | 527.9 | 94.07 | 1.3 | 35.93 |

## Tensile

Figure 3 shows typical stress-strain results for the three panels tested under uni-axial tension. Each curve demonstrates the highly nonlinear behavior during deformation. It is theorized that this non-linearity is due to delamination followed by yarn rotation, aligning the fibers with the applied force. From these curves it is unclear as to how effective either treatment is in limiting fiber mobility. From Table 1, we can see that although the plasma treatment results in an increase in both modulus and tensile yield strength, the addition of SiO$_2$ does not affect modulus, but does further increase the yield strength. The ultimate tensile strength of the composite does not appear to be greatly affected by the treatments.

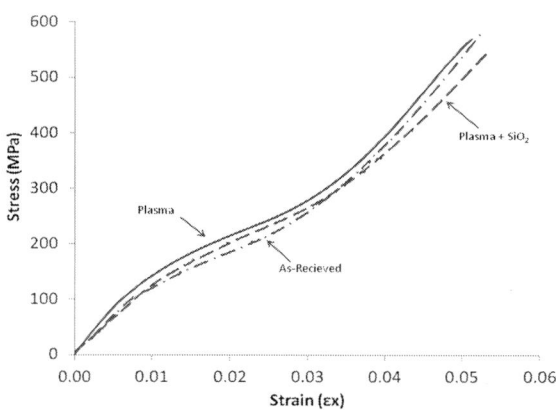

**Figure 3.** Typical stress-strain curves from tension tests of as-received and treated Kevlar composites

Figure 4 shows optical photographs of two specimens tested under tension. Some observations can be made by examining the damage in the tested specimens. The failure of the specimens often occurred within the gage region although each layer did not fail at the same location. A second observation is the significant decrease in the amount of delamination damage in the plasma treated specimens. This last observation is a clear indication of an increase in the inter-layer bonding due to the surface treatments.

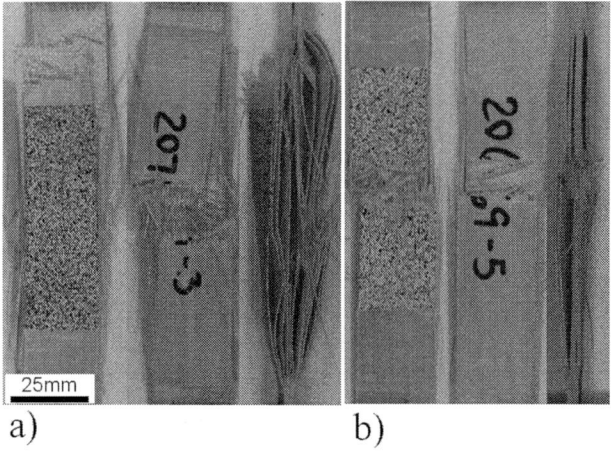

**Figure 4.** Optical micrographs of specimens tested in tension.
   a) Untreated Kevlar specimen showing large delamination damage, and
   b) Plasma + SiO2 treated specimen with reduced damage. Both Specimens show failure within the gage region.

## Compression

The typical mode of failure during compression testing is the formation of a kink-band which results from microbucking of the fibers. This behavior is more pronounced in woven composites due to the undulation of the yarns. During the DIC strain computation, this mode of failure was visible in specimens from each panel. Representative curves for the compressive stress-strain response of the Kevlar-SC15 composite panels are shown in Figure 5. From this figure it is clear that treatment of the Kevlar resulted in an increase in the in-plane compression strength. Averages from the results are given in Table 1. Both methods of plasma treatment resulted in an increase in strength. While treatment with plasma and $SiO_2$ resulted in a more then 35% increase in strength, this is still much lower than the in-plane tensile strength.

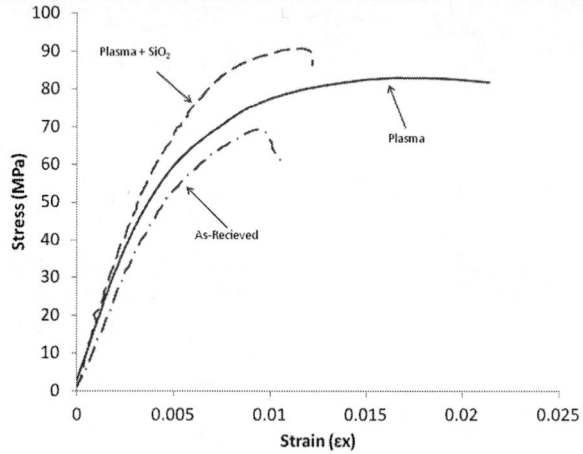

**Figure 1.** Typical stress-strain curves of as-received and treated Kevlar composites loaded in compression.

## Shear

Typical shear response curves for the three panels are shown in Figure 6. From this data the shear modulus and shear strength for each panel were calculated. The curves show a constant 'hardening response' during post-yield deformation which is typical for woven composites. This response is due to the rotation of the yarns which aligns the fibers with the principle strains. The results are summarized in Table 1. It was found that plasma treatment alone had only a slight increase in strength while decreasing the shear modulus. The addition of $SiO_2$ resulted in a significant increase in strength and slight increase in modulus.

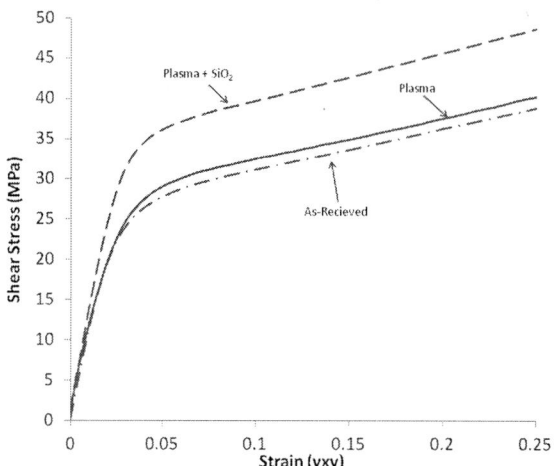

**Figure 2.** Typical shear stress-strain curves of Kevlar composites tested using v-notch beam method

## Conclusions

Results from the mechanical tests clearly indicate a significant change in the response of Kevlar-SC15 composites when the woven fabrics were subjected to surface enhancements using atmospheric plasma treatment. Plasma modification with $SiO_2$ was shown to enhance the mechanical properties including tensile yield, compressive and shear strength by as much as 35%. Observations of the induced damage indicate that surface modification using plasma may reduce the delamination and increase interfacial bond strength. Further analysis of the failure surface is necessary to characterize the mode of interface failure. Further research is needed to fully understand the effects of plasma treatments and to optimize techniques to improved performance and maximize the available strength of Kevlar fibers.

## References

1. J. R. Brown, Z. Mathys, "Reinforcement and Matrix Effects on the Combustion Properties of Glass Reinforced Polymer Composites," *Composites Part A: Applied Science and Manufacturing, 28A* (1997), 675–681.

2. P. Lee-Sullivan, K.S. Chian, C.Y. Yue, "Effects of Bromination and Hydrolosis Treatment on the Morphology and Tensile Properties of Kevlar-29 Fibers," *Journal of Materials Science Letters,* 13 (5) (1994), 305-309.

3. J. Zhao, "Effect of Surface Treatment on the Structure and Properties of para-Aramid Fibers by Phosphoric Acid," *Fibers and Polymers,* 14 (1) (2012), 59-64

4. A. S. Vasconcellos, J. A .P. Oliveira, R. Baumhardt-Neto, "Adhesion of Polypropylene Treated With Nitric And Sulfuric Acid," *European Polymer Journal,* 33 (1997), 1731–1734.

5. Y. Zhang, Z. Jiang, Y. Huang, Q. Li, "The Modification of Kevlar Fibers in Coupling Agents by γ-ray Co-irradiation," *Fibers and Polymers,* 12 (8) (2011), 1014-1020.

6. E. Kim, J. Jang, "Surface Modification of Meta-aramid Films by UV/ozone Irradiation", *Fibers and Polymers,* 11 (5) (2010), 677-682.

7. J. Breuer, S. Metev, G. Sepold, "Photolytic Surface Modification of Polymers With UV-Laser Radiation," *Journal of Adhesion Science and Technology,* 9 (1995), 351–363.

8. D. Zhang, Q. Sun, L. C. Wadsworth, "Mechanism of Corona Treatment On Polyolefin Films," *Polymer Engineering & Science,* 38 (1998), 965–970.

9. S. Farris, S. Pozzoli, P. Biagioni, L. Duó, S. Mancinelli, L. Piergiovanni, "The Fundamentals of Flame Treatment for the Surface Activation of Polyolefin Polymers – A Review," *Polymer* 51 (2010), 3591–3605.

10. C. Y. Yue, G. X. Sui, H.C. Looi, "Effects of Heat Treatment on the Mechanical Properties of Kevlar-29 Fibre," *Composites Science and Technology,* 60 (2000), 420-427.

11. R. E. Allred, E. W. Merrill, and D. K. Roylance, *Handbook of Composites,* (New York, NY, Van Nostrand Reinhold, 1982) p. 333.

12. Z. Dong, J. M. Manimala, C. T. Sun, "Mechanical Behavior of Silica Nanoparticle-Imprengated Kevlar Fabrics," *Journal of Mechanics of Materials and Structures,* 5 (4) (2010), 529-548.

13. W. Chen, X. Qian, X. He, Z. Liu, J. Liu, "Surface Modification of Kevlar by Grafting Carbon Nanotubes," *Journal of Applied Polymer Science,* 123 (2012), 1983-1990.

14. A. Bogaerts, E. Neyts, R. Gijbels, J. van der Mullen, "Gas Discharge Plasmas and Their Applications," *Spectrochimica Acta Part B: Atomic Spectroscopy,* 57 ( 2002), 609–658.

15. C. Tendero, C. Tixier, P. Tristant, J. Desmaison, P. Leprince, "Atmospheric Pressure Plasmas: A Review," *Spectrochimica Acta Part B: Atomic Spectroscopy,* 61 ( 2006), 2–30.

16. S. Bhattacharya, R. K. Singh, S. Mandal, A. Ghosh, S. Bok, V. Korampally, K. Gangopadhyay, S. Gangopadhyay, "Plasma Modification of Polymer Surfaces and Their Utility in Building Biomedical Microdevices," *Journal of Adhesion Science and Technology,* 24 (2010), 2707–2739.

17. A. Baldan, "Adhesively-Bonded Joints and Repairs in Metallic Alloys, Polymers and Composite Materials : Adhesives, Adhesion Theories and Surface Pretreatment," *Journal of Materials Science,* 39 (2004), 1–49.

18. M. J. Shenton, G. C. Stevens, "Surface Modification of Polymer Surfaces : Atmospheric Plasma vs. Vacuum Plasma Treatments," *Journal of Physics D: Applied Physics,* 34 (2001), 2761–2768.

19. D.D. Pappas, A. A. Bujanda, J. A. Orlicki, R. E. Jensen, "Chemical and Morphological Modification Of Polymers Under a Helium–Oxygen Dielectric Barrier Discharge," *Surface and Coatings Technology,* 203 ( 2008), 830–834.

20. D. Pappas, A. Bujanda, J. Demaree, J. Hirvonen, W. Kosik, R. Jensen, S. Mcknight, "Surface Modification of Polyamide Fibers and Films Using Atmospheric Plasmas," *Surface and Coatings Technology,* 201 ( 2006), 4384–4388.

21. Michael A. Sutton, Jean-Jose Orteu, and Huber W. Shreier, *Image Correlation for Shape, Motion and Deformation Measurements* (New York, NY: Springer Science+Business Media L.L.C. 2009), 321

22. T.C. Chu, W.F. Ranson, M.A. Sutton, W.H. Peters, "Applications of digital-image-correlation techniques to experimental mechanics" *Experimental Mechanics*,25 (1985) 232-244.

23. P. Moy, C.A Gunnarsson, and J. Tzeng. "Using Digital Image Correlation to Acquire Full-field Displacements and Strains in Carbon Fiber Tensile Tow Experiments and Modified Iosipescu Shear Specimens" (Paper presented at the 53rd International SAMPE Symposium and Exhibition, Long Beach, CA, 20 May 2008), 9.

24. Standard Test Method for Tensile Properties of Polymer Matrix Composite Materials. ASTM International, West Conshohocken, PA, www.astm.org; ASTM Standard D3039/D3039M, 2005

25. Standard Test Method for Compressive Properties of Polymer Matrix Composite Materials Using a Combined Loading Compression (CLC) Test Fixture. ASTM International, West Conshohocken, PA, www.astm.org; ASTM Standard D5379/D5379M, 2005

26. Standard Test Method for Shear Properties of Composite Materials by the V-Notched Beam Method. ASTM International, West Conshohocken, PA, www.astm.org; ASTM Standard D5379/D5379M, 2005

27. Standard Test Method for Shear Properties of Composite Materials by the V-Notched Rail Shear Method. ASTM International, West Conshohocken, PA, www.astm.org; ASTM Standard D7078/D7078M, 2005

Advanced Composites for Aerospace, Marine, and Land Applications
Edited by: Tomoko Sano, T.S. Srivatsan, and Michael W. Peretti
TMS (The Minerals, Metals & Materials Society), 2014

# FORMING LIMIT DIAGRAM OF STEEL/POLYMER/STEEL SANDWICH SYSTEMS FOR THE AUTOMOTIVE INDUSTRY

Mohamed Harhash[1], Adele Carrado[2] and Heinz Palkowski[1,*]

[1] *Metal Forming and Processing, Institute of Metallurgy*
Clausthal University of Technology
Robert-Koch-Str. 42, 38678 Clausthal-Zellerfeld, **Germany**

[2] *Institut de Physique et Chimie des Materiaux de Strasbourg, IPCMS,*
UMR 7504 ULP-CNRS, 23 rue du Loess, BP 43, 67034 Strasbourg cedex 2, **France**
*corresponding author: heinz.palkowski@tu-clausthal.de

## Abstract

Steel/polymer/steel sandwich materials (SMs) are considered an innovative substitute to the commercial steel sheets in the automotive industry due to weight-saving potential and enhanced damping properties. Deep-drawing steel and thermoplastic core are used as the skin and core sheets, respectively. The mono-materials and SMs were characterized via tensile test, deep drawing and flow limit curves (FLC) determination. The mechanical properties of the tested SMs showed a good matching with the mixture rule regardless the skin/core sheet thickness. Varying the skin/core sheet thickness has a remarkable effect on the thinning behavior under deep drawing; the core thickness increases the cracking probability. In case of using SMs with different skin-thickness, it is better to position the thin skin sheet in contact with the punch. The core thickness exhibited no significant effect on the FLC results. The thicker-core SMs are subjected to failure at lower strains in the stretching region of the FLC.

**Keywords:** steel/polymer/steel sandwich, forming, deep drawing, FLC, mechanical properties

## Introduction

Due to the demands for energy saving and better environmental impact of the mobile vehicles, there is a need to develop light-weight sheets. Steel/polymer/steel sandwich materials, SMs, provide an innovative substitute for the used commercial sheets. Currently, there are some ongoing projects in research and development of several car manufacturers for instance the SuperLightCar project [1].

In the current study, thermoplastic core is employed. The main advantages of using thermoplastic core are; good forming potential compared to the fiber metal laminates (FML) such as CARAL or GLARE which contain a relatively brittle core [2], in addition to the possibility on an integrated automated production line [3]. Characterizing the forming behavior is one of the most critical questions if the sandwich sheet is aimed for mobile applications [4, 5].

The importance of the forming limit curve (FLC) is to offer a chance to determine process limitations in sheet metal forming and is used in the estimation of stamping characteristics of sheet metal materials [6]. The forming potential of SMs based on Al alloy as the skin sheet was studied under deep drawing and the FLC curve was determined as a function of the core thickness. Due to the excessive tensile stresses exerted on the outer skin of the sandwich, the core thickness has a negative effect on the limiting draw ratio ($\beta$) [7, 8]. The strain distribution for the FLC curve is better for the AA5052/polyethylene/AA5052 than the monolithic AA5052 sheet. Moreover, it was found that increasing the core thickness from 0.5 to 2.0 mm has relatively a positive impact on the FLC and the dome height [9].

The aim of this paper is to investigate the effect of varied core thicknesses on the mechanical properties in terms of the elastic modulus, yield and ultimate tensile strength. Moreover, the dependency of the forming behavior on the core/skin sheet thickness varied is determined using deep drawing in addition to FLC curve determination.

## Experimental Work

In this study, SMs was composed of deep drawing steel grade 316L with a skin sheet thickness of 0.49 and 0.24 mm, while the core sheet was a PP-PE copolymer foil of 0.3, 0.6, 1.0 and 2.0 mm thickness. The SMs were produced in the lab via a two-step roll-bonding process using the compatible adhesive agent Köratec FL 201. The detailed production procedure is described and illustrated in [10]. The chemical composition for the polymeric core and the adhesive agent is listed in a previous thesis [11]. After fabrication, the SMs were cut and marked by electrochemical etching for the subsequent photogrammetical analysis.

Table 1 gives an overview of the performed test-plan for the current study. The main prospective of this paper is the mechanical and forming potential characterization. The effect of the core thickness on the mechanical properties was characterized via tensile test in addition to ascertain the applicability of the rule of mixture to be applied on other SMs combinations.

Table 1. Variation of layer thicknesses and performed tests; Tests: "o" no, "x" yes, "*" the side in contact with the punch during deep drawing.

| Material Notation | Thickness [mm] | | $f_{core}=$ | Tensile test | Deep drawing | FLC |
|---|---|---|---|---|---|---|
| | Skin | Core | $t_{core}/t_{SMs}$ | | | |
| St. 0.49 | 0.49 | 0 | 0 | x | x | x |
| 0.49/0.3/0.49 | 0.49 | 0.3 | 0.23 | x | x | o |
| 0.49/0.6/0.49 | 0.49 | 0.6 | 0.38 | x | x | x |
| 0.49/1.0/0.49 | 0.49 | 1.0 | 0.51 | o | x | o |
| 0.49/2.0/0.49 | 0.49 | 2.0 | 0.67 | x | x | x |
| St. 0.24 | 0.24 | 0 | 0 | x | x | x |
| 0.24/0.3/0.24 | 0.24 | 0.3 | 0.38 | x | o | x |
| 0.24/0.6/0.24 | 0.24 | 0.6 | 0.56 | x | x | o |
| 0.24/2.0/0.24 | 0.24 | 2.0 | 0.81 | x | o | o |
| 0.24*/0.6/0.49 | | 0.6 | 0.45 | o | x | o |
| 0.49*/0.6/0.24 | | 0.6 | 0.45 | o | x | o |

In order to evaluate the forming behavior of the SMs, the following tests were carried out;

- Deep drawing, using a cylindrical flat punch of ∅ 33 mm ($D_p$) and a blank diameter of ∅ 68 mm ($D_b$) reaching a limiting drawing ratio β = $D_b/D_p$ = 2.06. The blank holding force was optimized to avoid cracking and wrinkling defects. Based on the layer thicknesses, it ranged between 7-9 kN. Moreover, a 0.2 mm thick polymeric foil was used as a lubricant additionally to preserve the marks used for the photogrammetrical analysis afterwards. The deep drawing test configuration is illustrated in Figure 1-b. The forming behavior via deep drawing was investigated on SMs at different core thicknesses in addition to different setting conditions at different skin sheet thicknesses in respect to the punch as shown in Table 1.

- FLC evaluation was carried out for the monolithic steel sheets of both thicknesses in addition to three other SMs combinations, following DIN 12004-2 [12], as shown in Table 1. A semispherical punch of ∅ 75 mm and blank of ∅ 180 mm were used. The drawing speed and the applied blank holder force were 1.5 mm/s and 100 kN, respectively. In order to stimulate different failure strains the width of the samples was changed in seven steps as following: 180, 140, 100, 90, 80, 50 and 20 mm. The FLC test settings and sample geometry are shown in Figure 1-a and -c, respectively. A lubricant configuration of a 0.6 mm polymeric foil covered with a high performance grease was used. The major and minor strains were determined using photogrammetrical analysis with 15 frames per second. The FLC curves were determined at the necking, just before cracking.

## Results and Discussion

### Mechanical properties

Figure 2-a and –b show stress-strain curves for the monolithic steel sheets (St. 0.49 and St.1.24 mm) and the SMs with different core thicknesses. It can be clearly observed that varying the core thickness has a systematic impact on the mechanical properties. With increasing the core thickness, the mechanical properties - such as the elastic modulus, yield and ultimate tensile strength - decrease in accordance with the rule of mixture. The verification of the rule of mixture based on the mechanical properties of the mono-materials at different volume fractions of the core are illustrated in Figure 2-c and –d. There is a very good agreement between the measured and the estimated values. In this regard, it can be stated that the rule can be applied for such skin/core thickness combinations.

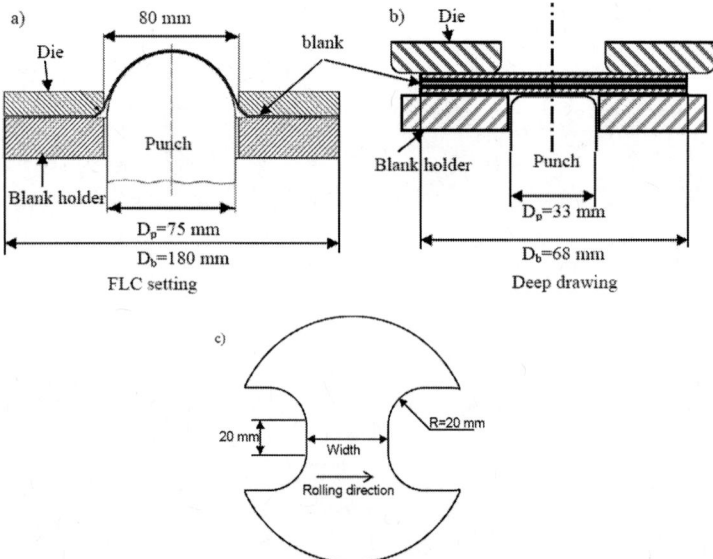

**Figure 1.**  FLC and deep drawing testing setting; a) and b), respectively, c) One test specimen geometry for the FLC test

It can be observed as well that there is no remarkable change in elongation at failure. This can be attributed to the dominance of the skin steel sheets conveying the load and fail first. However, the polymeric core shows further straining after the failure of the skin sheets at low level of stress. It can be stated that no delamination occurs during tensile testing. Varying the skin/core thickness combination at constant core volume fraction shows no difference of the mechanical properties. This conclusion can be shown comparing Figure 2-c and –d at $f_{core}$ of 0.38 for the two SMs 0.49/0.6/.049 and 0.24/0.3/0.24.

## Deep drawing

The effect of the core thickness on the drawing force at constant skin thickness of 0.49 mm is illustrated in Figure 3-a. The drawing force was evaluated stepwise for a drawing cup depth of 5, 10, 15, 20 mm and until complete drawing. It can be observed that there is no remarkable effect of the core thickness on the drawing force for the core thicknesses of 0.3, and 1.0 mm at the five tested steps. However, for the thicker core of 2.0 mm, the drawing force decreases. This can be attributed to low force required to form the thick soft core to reach the same cup depth for the other SMs. Additionally, The SMs 0.49/2.0/0.49 does not completely drawn but cracks at a cup depth of 20 mm.

The effect of varying the skin thickness and the setting conditions in respect to the punch at constant core thickness of 0.6 mm is shown in Figure 3-b. When both skin sheets are thick in SMs 0.49/0.6/0.49, the maximum force is reached in comparison to SMs 0.24/0.6/0.24. It can be observed that the SMs 0.24*/0.6/0.49 shows relatively higher drawing force than SMs 0.49*/0.6/0.24 where the outer skin is the thicker one. In addition, the thinner sheet showed cracking at the sidewall in SMs 0.49*/0.6/0.24. It can be stated that the drawing force is mainly dependent on the skin sheet thickness and its relation the punch.

Although there is no remarkable difference of the drawing force, the thickness strain distribution is affected significantly by the core thickness at constant skin thickness as shown in Figure 4. Figure 4-a shows the real drawn cups at the final step and their corresponding 3D images built by the GOM software. The 3D images represents the thickness strain in log. Figure 4-b shows thickness strain distribution along a section in the rolling direction for the different SMs at constant skin thickness at the drawn stage. Figure 4-c represents the maximum thickness strain at the punch rounding (the bottom/sidewall transition) at each stage of the drawing process.

It can be observed that with increasing the core thickness the thickness strain at the punch rounding increases as a result of the tensile strains acting on the outer surface accelerating failure by cracking. As a result, the SMs with 2.0 mm core is subjected to cracking. During the drawing process of the three-layered structure, the punch supports and stabilizes the inner skin transferring the load to the outer skin through the core. Due to the existed thick and elastic core, the outer skin is subjected to higher thickness strain at the punch rounding leading to cracking. It was observed that there is small wrinkling developing at the edges of the SMs just at the final step of drawing as shown in Figure 4-a.

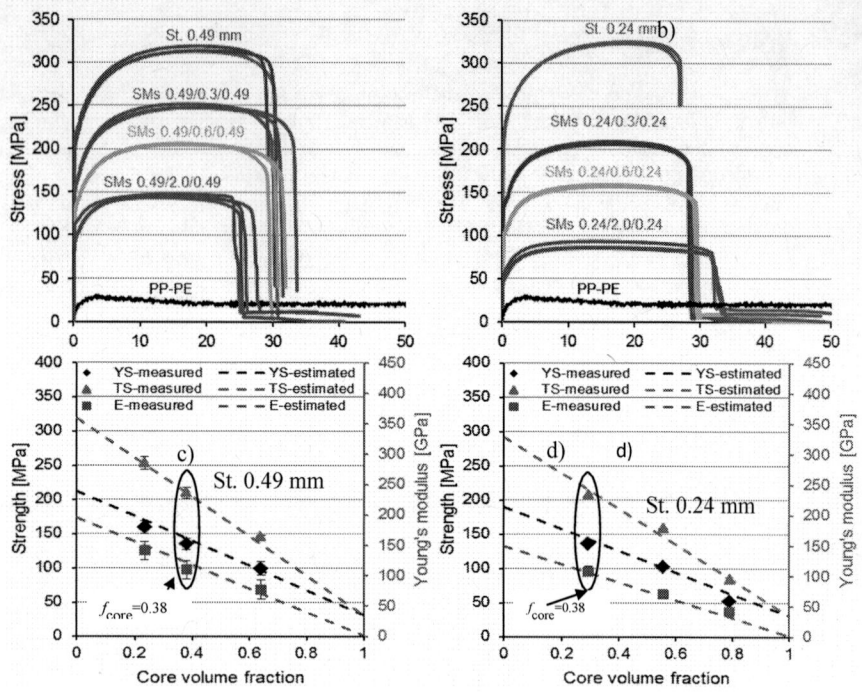

**Figure 2.** Stress-strain diagrams showing the effect of the core thickness on the mechanical properties (a and b) and their verification with estimated values according to the rule of mixture for St. 0.49 and St. 0.24 (c and d).

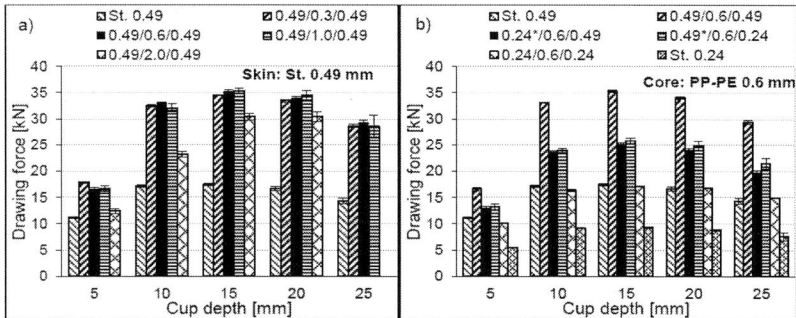

**Figure 3.** Drawing force vs. cup depth (flat cylindrical punch ∅ 33 mm:
(a) Varying the core thickness at constant skin sheet thickness of 0.49 mm
(b) Varying the skin sheet setting condition at core sheet thickness 0.6 mm
* refers to the skin sheet in contact with the punch

The effect varying the setting conditions is shown in Figure 5-a. When the thicker steel sheet is in contact with the punch, the thinner sheet (0.24 mm) is subjected to excessive stretching until reaching the failure conditions at the sidewall. However, when the setting is reversed, the outer thicker sheet does not crack such. Figure 5-b shows the maximum thickness strain at the punch rounding for different SMs and skin sheets. The thicker is the outer skin, the higher straining it shows. It can be stated that the best setting condition for different skin thicknesses is to position the thicker sheet as the outer one to extend the forming limit if possibe.

**FLC determination**

FLC was determined at the necking stage following DIN 12004-2 [12], just before cracking as shown in Figure 6-a for SMs 0.49/2.0/0.49 as an example. The FLC was determined by plotting all minor-major strain ($\varepsilon_2$-$\varepsilon_1$) points of the detected surface area for all the samples at different sample widths. The top points (major strain) are connected giving the FLC curve. It can be observed that the failure limits are between the guidelines $\varepsilon_1 = \varepsilon_2$ in the stretching zone (right) and $\varepsilon_1 = -2\varepsilon_2$ in the drawing zone (left). Comparing the FLC for the thin and thick skin sheets, Figure 6-b shows that St. 0.49 has an improved limit curve at necking compared to St. 0.24.

The effect of the core thickness on the FLC is illustrated in Figure 7-a. It can be clearly observed that there is no significant difference between the monolithic steel sheet St. 0.49 thickness and the two other sandwiches of different core thickness (0.6 and 2.0 mm). This can be attributed to the same behavior shown by the skin sheet because it is always the detected outer sheet. However, there is a remarkable influence of the core thickness in the area of stretching. With increasing core thickness failure of the SMs occurs earlier. Moreover, there was some increment of the drawing force with increasing core thickness as shown in Figure 8. In case of comparing the St. 24 with SMs 0.24/0.3/0.24, no significant difference can be observed as shown in Figure 7-b. It can be stated that varying the core thickness has no significant effect on the forming limit curve however the FLC is mainly influenced by the skin sheet thickness.

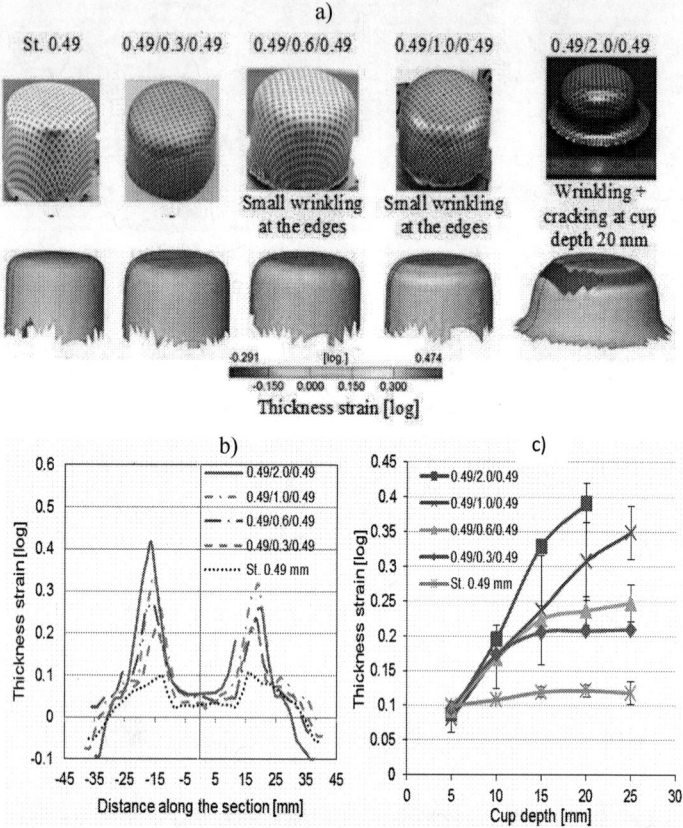

**Figure 4.** The effect of the core thickness on the thickness strain distribution in terms of:
(a) Real cup and their 3D images,
(b) Thickness strain profile along a section in the rolling direction for the drawn cup,
(c) The cup depth at varying core thickness.

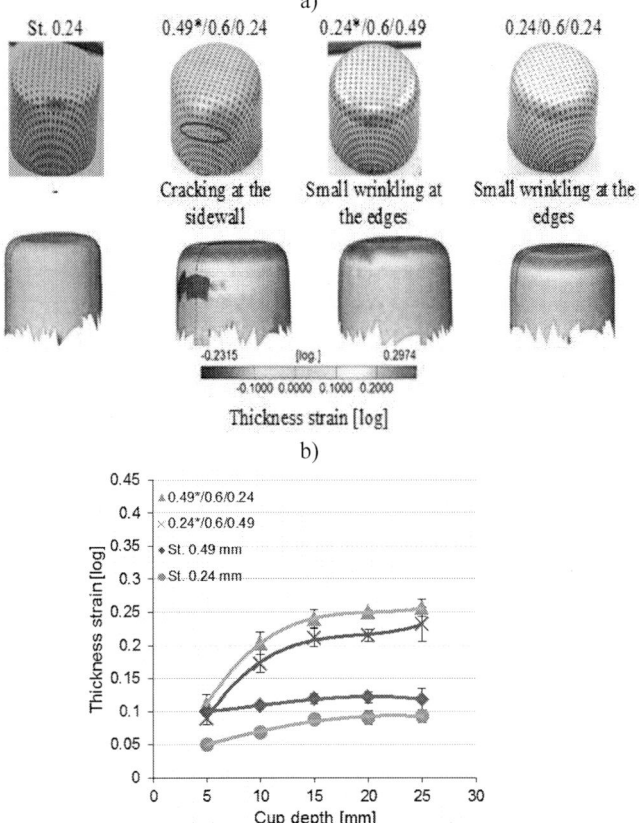

Figure 5. The effect of the setting conditions on the thickness strain distribution in terms of:
(a) Photos of cups and their 3D srain images
(b) Cup depth at varying setting conditions.

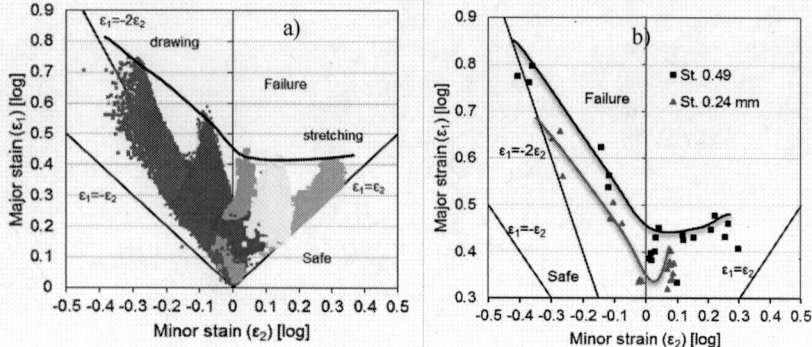

**Figure 6.** FLC at the necking stage for SMs 0.49/2.0/0.49 (a), b) FLC for the steel sheets St. 0.49 and St. 0.24.

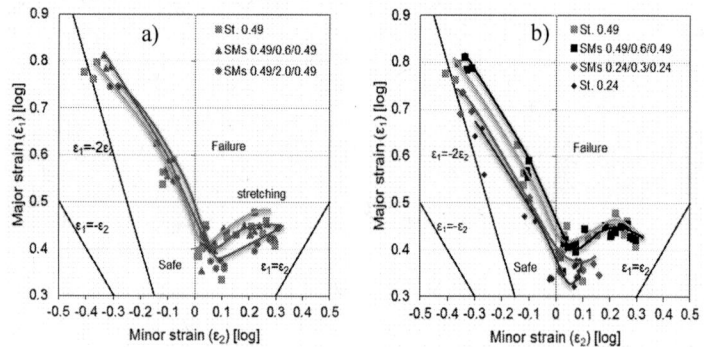

**Figure 7** (a) Effect of the core thickness on the FLC for St. 0.49 skin sheet
(b) Comparing the FLC for SMs of the same core volume fraction at different skin/core thicknesses.

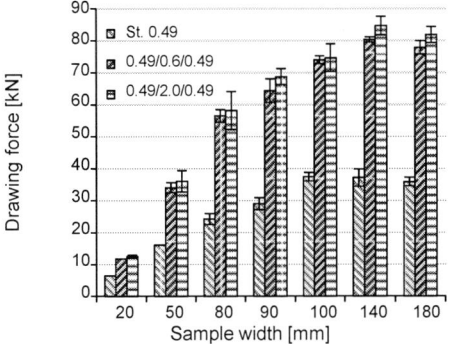

**Figure 8.** Relation between drawing force and core thicknesses for different web widths of the samples.

## Conclusions

The investigation clearly shows that

1. The core thickness has a remarkable effect on the tensile properties. The rule of mixture is applicable for the tested sandwich combinations and can be applied for other combinations.

2. Variation of the skin/core thickness has a significant effect of the forming behavior; the thicker is the core, the earlier cracking under deep drawing and stretching conditions occurs.

3. In case of varied skin thickness in the same sandwich, the best setting condition is to position the thinner skin in contact with the forming tool to avoid earlier failure.

4. The FLC is mainly affected by the shin sheet thickness. The thicker sheet exhibited better forming limit. The core thickness exhibited no significant effect on the FLC results in the range of the investigated thicknesses. However, the thicker-core SMs tend to fail at lower strains in the stretching region of the FLC.

### Acknowledgement

The authors would like to thank the German Research Foundation (DFG) for the financial support in addition to DAAD for supporting the first author with a scholarship within the GERLS program.

## References

1. Goede, M. SuperLIGHT-Car project – An integrated research approach for lightweight car body innovations. in International Conference of Innovative Developments for Lightweight Vehicle Structures. 2009. Wolfsburg, Germany.

2. Sinmazçelik, T., E. Avcu, M.Ö. Bora and O. Çoban, A review: Fibre metal laminates, background, bonding types and applied test methods. *Materials & Design*, 2011. 32(7): p. 3671-3685.

3. ThyssenKrupp-Stahl, Bondal: Composite Material with structure-brone sound damping properties 2009: Germany.

4. Engel, B. and J. Buhl, Metal Forming of Vibration-Damping Composite Sheets. *steel research international*, 2011. 82(6): p. 626-631.

5. Engel, B. and J. Buhl, Forming of Sandwich Sheets Considering Changing Damping Properties, in Metal Forming - Process, Tools, Design, M. Kazeminezhad, Editor. 2012.

6. GOM-ARAMIS, Determination of Process Limitations in Sheet Metal Forming - Forming Limit Diagram, in Material Testing / Material Properties, GOM-mbH, Editor. 2009: Germany.

7. Parsa, M.H., M. Ettehad, P.H. Matin and S.N. Al Ahkami, Experimental and Numerical Determination of Limiting Drawing Ratio of Al3105-Polypropylene-Al3105 Sandwich Sheets. *Journal of Engineering Materials and Technology*, 2010. 132(3): p. 031004-031004.

8. Marumo, Y., H. Saiki and L. Ruan, Effect of sheet thickness on deep drawing of metal foils. *Journal of Achievements in Materials and Manufacturing Engineering*, 2007. 20(1-2).

9. Liu, J.-g. and W. Xue, Formability of AA5052/polyethylene/AA5052 sandwich sheets. *Transactions of Nonferrous Metals Society of China*, 2013. 23(4): p. 964-969.

10. Sokolova, O., A. Carradó and H. Palkowski, Production of Customized High-Strength Hybrid Sandwich Structures. *Advanced Materials Research*, 2010. 137: p. 81-128.

11. Sokolova, O., Dissertation thesis "Study of metal/polymer/metal hybrid sandwich composites for the automotive industry", in Metal Forming and Processing. 2012, TU Clausthal: Germany. p. 159.

12. DIN-12004-2, Determination of forming-limit curves in the laboratory, in Metallic materials – Sheet and strip - Determination of forming limit curves. 2008, Beuth Verlag GmbH: Berlin.

# The Wettability of TiC$_x$ by Titanium-Aluminum Alloys at 1758 K

Xuyang Liu, Xuewei Lv [*], Chenguang Bai, Chunxin Li

College of Materials Science and Engineering
**Chongqing University**,
Chongqing 400044, China
Email: liuxuyangcqu@163.com.
lvxuewei@163.com

## Abstract

The wetting of ceramic by liquid metals is one of the most important phenomena during processing metal matrix composite material. The wettability of TiCx substrate by the liquid Ti-Al alloys with 40, 50, 60, 70 and 80 wt. % Al was studied using the sessile drop method at 1758 K in an Ar atmosphere. The results show that the initial angle decrease from 110° to 80° with increases of Al, and the final equilibrium angle are lower than 10°and slightly depend on the contents of Al. The diffusion of carbon or titanium between Ti-Al alloy and TiCx substrate and interfacial reaction promote the wettability.

**Key words:** wetting, Ti-Al alloys, titanium carbide

## Introduction

Wetting phenomena is considered an important phenomenon during processing for metallic matrix composites between metal and ceramic [1-2]. It is not only depends on thermo physical property of system, but has relationship with interface chemical reaction. It is no doubt that interface behavior between solid and liquid is the most critical issues to be solved, which determine the comprehensive performance of material directly, such as strength, stiffness, toughness and so on.

The wettability between metal and ceramic has been reported so much [3-15], but very little work has concerned with the wetting of Ti-Al alloy on TiC substrate. Based on the deficiency about wettability of Ti-Al alloy on TiC and meaningfulness for the following preparation of composite material, the aim of this investigation was to study the wetting behavior of Ti-Al alloys on TiC as a function of chemical composition and clarify the mechanism of the wetting in this system.

## Experimental

The raw materials used in wettability study were TiCx substrates (20 mm*15 mm*3 mm) with 99.5% density and Ti-Al alloys with 40, 50, 60, 70 and 80 wt. % Al. The substrates were prepared by hot pressing under vacuum in a graphite furnace at 1800 °C using a constant pressure of 30 M Pa for 30 min. The substrates surface were polished by 3.5μm、2.5μm、1.5μm、0.5μm diamond paste for the polish in sequence and cleaned in acetone three times to remove impurities by Q2200E ultrasonic cleaner before each wetting experiment. The roughness of substance surface yielded during polish process were detected by Dektek with a 20 nm-25 nm Ra. TiCx has a wide non-stoichiometry from x=0.47 to 0.98. According to the relationship between the lattice parameter and C/Ti composition, the TiCx in this study was determined to be 0.78.

Pure (99.6%) Ti and high-pure (99.95%) Al was used to prepare Ti-Al alloy by powder metallurgy method. In order to make a more uniform composition, Ti and Al powder were mixed 10 min by high-energy ball milling under rotate speed of 200 r/ min. Then the mixture was uniaxial pressed in a hydraulic press with 20 M Pa to make piece with 15 mm in diameter and 6 mm in thickness. The pieces were sintered 6 h at 1100 °C in a vacuum of ~0.07 M Pa. After sintering, the alloys were cut in cubic shape (~0.03 g) and cleared in the same way with TiC substrates.

The wetting experiment was carried out in improved sessile drop equipment. The apparatus mainly consists of a heating system with a tantalum cylindrical heater, an evacuating pump system with a rotary pump and a molecular pump, a purification of the argon gas system with magnesium, a lifting system with high-pure alumina supporter inside the chamber, a WRe5/26 type thermocouple, a data acquisition system, He–Ne lasers system and high-resolution (2000*1312 pixels) digital cameras. The TiC substrate was preplaced on alumina supporter in the center of the chamber and adjusted to a horizontal position. The Ti-Al alloy was placed in a stainless-steel tube with a flexible connector on the top of the dropping device outside the chamber, which connected an open alumina tube located in the right above TiC substrate.

The chamber was evacuated to 5 Pa first at room temperature by rotary pump and further evacuated to 5 * 10-4 Pa by molecular pump. The Ar gas (99.9999%) purified by passing through a magnesium containing furnace at 723 K, was introduced into chamber with the pressure of 0.8 atmosphere to mitigate Al evaporation. Then the chamber was heated to 1237 K at a heating rate of 20°C/min and to 1758 K at 10 °C/min. After temperature and pressure stabilized, the Ti-Al alloy was dropped from the flexible connector through stainless-steel tube and alumina tube and rested on the TiC substrate, which enabling the isothermal wetting. The change of drop profiles were recorded by high-resolution digital cameras at various time intervals until a steady state, which reflected the change of contact angle, base diameter and height directly.

When the wetting experiment finished, the samples were cooled to room temperature at a cooling rate of 15°C/min. The cooled samples were sectioned perpendicular to interface after embedding in resin and then polished for phase detection by X-ray diffraction (XRD, Rigak D/Max-2500) with Cu Kα radiation using a scanning speed of 1.2 °/min and scanning electron microscopy( SEM, Evo18, Carl Zeiss, Germany). The schematic diagram of wetting apparatus is showed in Fig. 1.

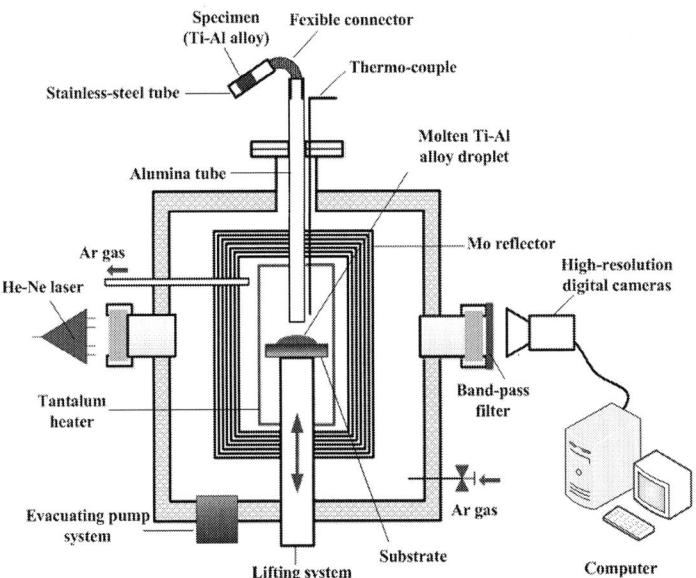

Figure 1.　　Schematic diagram for high-temperature wettability measurement.

## Results and Discussion

### The Preparation of Alloys

Figure 2 shows the XRD pattern for Ti-Al alloys with different Al contents prepared by power metallurgy. It can be seen that the main phases of Ti-Al alloy after sintering are TiAl, $TiAl_2$ and TiAl, $TiAl_2$ and $TiAl_3$, Al and $TiAl_3$, Al and $TiAl_3$ with Al content increasing respectively. The phase compositions presented in Figure 2 are corresponding with Ti-Al binary phase diagram.

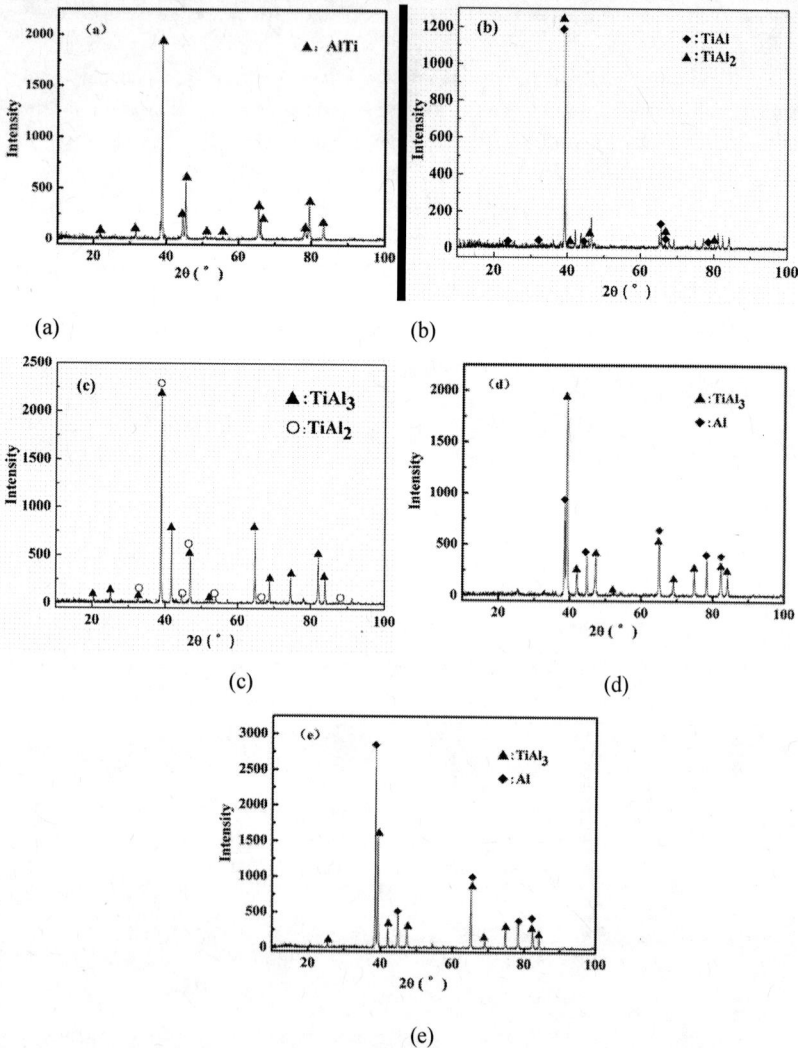

(a)

(b)

(c)

(d)

(e)

**Figure 2.** XRD pattern for Ti-Al alloys with different Al contents prepared by power metallurgy
(a) Ti-40 wt. % Al, (b) Ti-50 wt. % Al, (c) Ti-60 wt. % Al, (d) Ti-70 wt. % Al,
(e). Ti-80 wt. % Al

## Wetting Phenomena

Figure 3 is the variations in contact angle with time of Ti-Al alloy on TiC substrate during isothermal dwells at 1758 K. The main variation trend could be divided into three stages: (i) a rapidly decreasing stage where contact angle decreased quickly during several seconds; (ii) a spreading stage where the interfacial front advancing and contact angle decreased slowly; (iii) a final equilibrium stage where contact angle is almost unchanged with time.

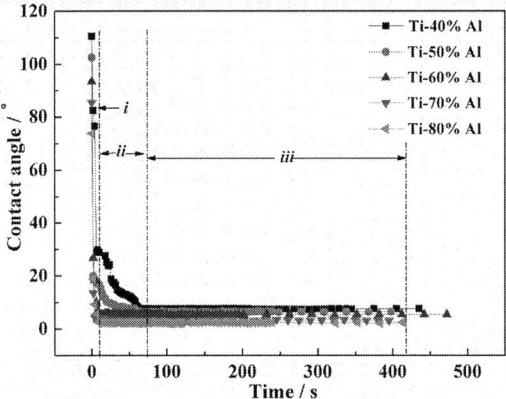

**Figure 3.** Kinetic spreading of contact angle with time during isothermal dwells at 1758 K for Ti-Al alloy/TiC system

Besides, it can be seen that initial contact angle is decreased with content of Al in Ti-Al alloy increasing and range of initial contact angle is between 110° -80°, which can illustrate that the wetting behavior of Ti-Al alloy with TiC substrate is partial wetting and the wetting can be improved with content of Al increasing and Ti decreasing. The variation of final contact angle shows the small relationship with composition of alloy.

## Interfacial characterization and Wetting Mechanism

Figure 4 shows the microstructures of the cross-sectional interface for Ti-40 wt. % Al alloys/TiC systems. It can be seen clearly that a layered microstructure had been formed, which indicated that chemical reactions occurred at the interface in the samples. The results of energy spectrum analysis show the main phase on the interface is $AlTi_2C$. According to the microstructures of the cross-sectional interface, it can be concluded that Ti-Al/ TiC is a reactive system.

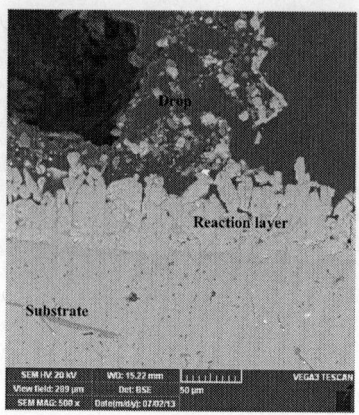

**Figure 4.** The microstructures of the cross-sectional interface for Ti-40 wt. % Al alloys/TiC systems

The thermodynamic reaction can be illustrated by Ti-Al-C ternary phase diagram. According to Ti-Al-C ternary phase diagram, the reaction process during wetting was inferred as follows: when Ti-Al alloys with TiC substrate is contacted, the chemical reaction on the interface of solid-liquid starts to form ternary carbides by the peritectic reaction. Then $Al_4C_3$ will be formed when the TiC concentrate is high.

$$L + TiC \to Ti_2AlC$$

$$L + TiC \to Al_4C_3 + TiC$$

Due to the limited kinetic and thermodynamic conditions, the carbon concentrate diffused from substrate to alloys is restricted. So there is no sufficient carbon to react with liquid to form $Al_4C_3$ phase. Besides, the formation of $Ti_2AlC$ phases on the interface can further hinder the contraction between Ti-Al alloy and TiC, thus led to there is no more carbon or titanium diffused into liquid. That's why the only $Ti_2AlC$ existed in the interface of solid-liquid rather than $Al_4C_3$ phase reported in literatures.

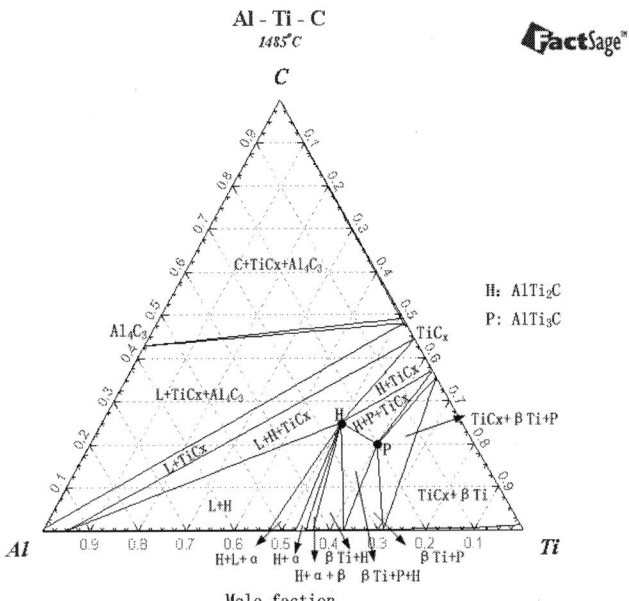

**Figure 5.** Ti-Al-C ternary phase diagram at isothermally of 1758 K

In metal/carbide systems, the spreading usually can be driven by dissolution, adsorption or reaction. Considering titanium carbide is a solid solution between titanium and carbon. The TiCx substrates were detected by XRD to illustrate whether the stoichiometry of TiCx changed after wetting. Due to the variation of detail of TiCx peaks of alloys are the similar, here Fig. 6 shows the detail of TiCx peaks before and after the wetting experiment at 1758 K of Ti-60 wt. % Al alloy. The results show that the position of TiCx peaks had shifted to higher angles in all of the samples, meaning that the smaller lattice parameter obtained when compared with the initial non-wetting substrate. The relationship between lattice parameter and carbon-titanium ratio was obtained from the references[16-17]. So it is indicated in Fig. 6 that carbon-titanium ratio of TiCx after wetting is lower than it before wetting. It can be illustrated the carbon in substrate diffused to alloy or more titanium in liquid diffused into substrate during wetting.

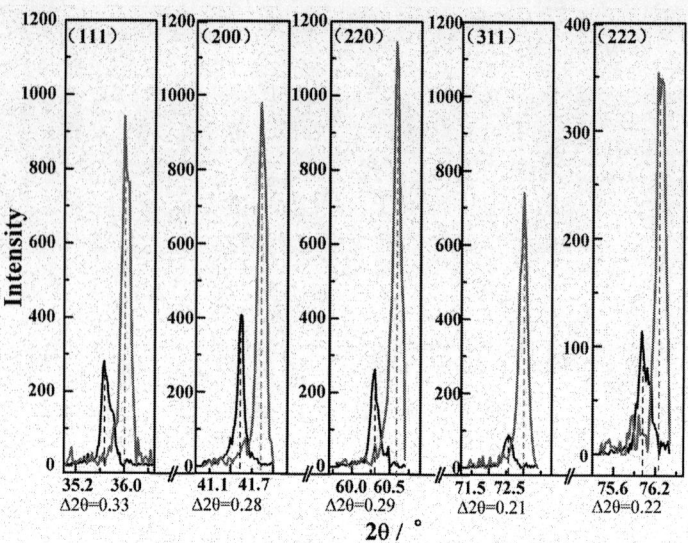

**Figure 6.** The TiC peaks before and after the wetting experiment at 1758K of Ti-60 wt. % Al alloy
(<u>Red line</u>: after wetting; <u>Black line</u>: before wetting)

## Conclusions

1. Ti-Al alloy/ TiC system is a reactive system, the main interface product is $Ti_2AlC$.

2. The initial angle decrease from 110° to 80° with increases of Al, and the final equilibrium angle are lower than 10° and slightly depend on the contents of Al.

3. The diffusion of carbon or titanium between Ti-Al alloy and TiCx substrate and interfacial reaction promote the wettability.

## References

[1] Contreras, A, VH Lopez, E Bedolla. Mg/TiC composites manufactured by pressureless melt infiltration[J]. *Scripta materialia*, 2004, 51 (3): 249-253.

[2] Zhang, Dan, Ping Shen, Laixin Shi, Qichuan Jiang. Wetting of B4C, TiC and graphite substrates by molten Mg[J]. *Materials Chemistry and Physics*, 2011, 130 (1): 665-671.

[3] Lin, Qiaoli, Ping Shen, Longlong Yang, Shenbao Jin, Qichuan Jiang. Wetting of TiC by molten Al at 1123–1323K[J]. *Acta materialia*, 2011, 59 (5): 1898-1911.

[4] Landry, K, S Kalogeropoulou, N Eustathopoulos. Wettability of carbon by aluminum and aluminum alloys[J]. *Materials Science and Engineering*: A, 1998, 254 (1): 99-111.

[5] Li, YL, TG Zhou. Achieving Al Melt/Carbon and Al-Ti Melts/Carbon Interfaces Wetting via Ultrasonic Couple Processing[J]. *Metallurgical and Materials Transactions A*, 2013: 1-7.

[6] Rhee, SK. Wetting of ceramics by liquid aluminum[J]. *Journal of the American Ceramic Society*, 1970, 53 (7): 386-389.

[7] Kononenko, VI, GP Shveikin, AL Sukhman, VI Lomovtsev, BV Mitrofanov. Chemical compatibility of titanium carbide with aluminum, gallium and indium melts[J]. *Poroshk. Metall.*, 1976, (9): 48-52.

[8] Muscat, Daniel, Robin AL Drew. Modeling the infiltration kinetics of molten aluminum into porous titanium carbide[J]. *Metallurgical and Materials Transactions A*, 1994, 25 (11): 2357-2370.

[9] Froumin, N, N Frage, M Polak, MP Dariel. Wettability and phase formation in the TiCx/Al systems[J]. *Scripta Materialia*, 1997, 37: 1263-6.

[10] Contreras, A, CA Leon, RAL Drew, E Bedolla. Wettability and spreading kinetics of Al and Mg on TiC[J]. *Scripta Materialia*, 2003, 48 (12): 1625-1630.

[11] Xiao, Ping, Brian Derby. Wetting of titanium nitride and titanium carbide by liquid metals[J]. *Acta Materialia*, 1996, 44 (1): 307-314.

[12] N Froumin, N Frage, M Polak, MP Dariel. Wetting phenomena in the TiC/(Cu–Al) system[J]. *Acta Materialia*, 2000, 48 (7): 1435-1441.

[13]   Frage, N, N Froumin, MP Dariel. Wetting of TiC by non-reactive liquid metals[J]. *Acta Materialia*, 2002, 50 (2): 237-245.

[14]   Zisman, Wm A. Relation of the equilibrium contact angle to liquid and solid constitution[J]. *Advanced Chem. Series,* 1964, 43 (1): 1-49

[15]   Leon, CA, VH Lopez, E Bedolla, RAL Drew. Wettability of TiC by commercial aluminum alloys[J]. *Journal of Materials Science*, 2002, 37 (16): 3509-3514.

[16]   Pierson, Hugh O. Handbook of Refractory Carbides & Nitrides: Properties, Characteristics, Processing and Apps[M]: Access Online via Elsevier, 1996.

[17]   Kerans, RJ, KS Mazdiyasni, R Ruh, HA Lipsitt. Solubility of Metals in Substoichiometric TiC1-x[J]. Journal of the American Ceramic Society, 1984, 67 (1): 34-38.

# AUTHOR INDEX
# Advanced Composites for Aerospace, Marine, and Land Applications

## A
Abad, M. ............................................. 173
Altamirano, A. ...................................... 23
Ares, A. .............................................. 115

## B
Back, C. .............................................. 173
Baeza, J. ............................................. 231
Baharvandi, H. ..................................... 55
Bai, C. ................................................ 255
Bai, J. ................................................. 101
Bakhshi, S. ........................................... 55
Banerjee, R. ......................................... 67
Bartoli, I. ............................................ 137
Bhandakkar, A. ............................... 3, 185
Bodkhe, S. ........................................... 35
Borhani, G. .......................................... 55
Borkar, T. ............................................ 67
Brown, D. ........................................... 127
Bujanda, A. ........................................ 231

## C
Cortés, V. ............................................. 23
Cuadra, J. .......................................... 137
Cui, L. ............................................... 127

## D
Dash, R. .............................................. 43
Deck, C. ............................................ 173

## E
Earthman, J. ....................................... 213

## F
Frazer, D. .......................................... 173
Freeman, C. ....................................... 157

## G
Ganguly, R. ......................................... 43
Garfias, E. ........................................... 23

Gopagoni, S. ....................................... 67

## H
Hao, S. .............................................. 127
Harhash, M. ....................................... 243
Hazeli, K. .......................................... 137
Hernández, E. ...................................... 23
Hosemann, P. ..................................... 173
Hwang, J. ............................................ 67

## J
Jiang, D. ............................................ 127
Joseph, P. ...................................... 81, 91
Ju, J. ................................................. 101

## K
Kamle, S. ............................................ 35
Kontsos, A. ........................................ 137

## L
Leblanc, B. ........................................ 157
Li, C. ................................................ 255
Liu, H. .............................................. 101
Liu, X. .............................................. 255
Lv, X. ............................................... 255

## M
Manu, K. ....................................... 81, 91
McWilliams, B. .................................. 203
Mehdikhani, B. .................................... 55
Montesano, J. .................................... 147

## P
Pai, B. .......................................... 81, 91
Palkowski, H. ..................................... 243
Pappas, D. ......................................... 231
Poveromo, S. ..................................... 213
Prasad, R. .................................... 3, 185

# R

Rajan, T. .................................................. 81, 91
Rajesh, P. ...................................................... 35
Rana, P. ......................................................... 43
Refugio, E. .................................................... 23
Ren, Y. ......................................................... 127
Resmi, V. ................................................. 81, 91
Rigby, D. ..................................................... 157
Rodriguez-Santiago, V. .............................. 231

# S

Sastry, S. ................................................. 3, 185
Saxe, P. ....................................................... 157
Schvezov, C. ............................................... 115
Singh, C. ..................................................... 147
Srivatsan, T. ............................................ 81, 91

# T

Tiley, J. ......................................................... 67

# V

Vanniaparambil, P. .................................... 137
Vázquez, L. ................................................... 23
Verma, V. ..................................................... 35
Vite, M. ........................................................ 23

# W

Walter, T. ................................................... 231
Wang, S. ..................................................... 127

# X

Xue, F. ........................................................ 101

# Y

Yen, C. ....................................................... 203
Yim, J. ........................................................ 231
Yu, C. ......................................................... 127

# Z

Zhou, J. ...................................................... 101

# SUBJECT INDEX
# Advanced Composites for Aerospace, Marine, and Land Applications

## 3
3D Reconstruction ......................................... 67

## A
Alkali Activator ............................................. 43
Aluminium Alloy ..................................... 81, 91
Aluminium Alloy AA6061 ............................. 3
Aluminium Fly Ash Composites ................ 185
Alumino-silicate ............................................ 91
AOD Slag ...................................................... 43
Atmospheric Plasma ................................... 231
Atomistic Simulation .................................. 157

## B
B$_4$C-Al ........................................................... 23

## C
Ceramic Preform ........................................... 91
Co-Al Ratio ................................................. 101
Composites .......................................... 23, 213
Corrosion ...................................................... 23
Corrosion Resistance .................................. 115

## D
Damage ....................................................... 137
Damage Mechanics ............................. 147, 185
Deep Drawing ............................................. 243
Digital Image Correlation ........................... 231

## F
Fatigue ........................................................ 137
FIB .............................................................. 173
Fiber-reinforced Composites ...................... 137
Finite Elements ........................................... 147
FLC ............................................................. 243
Fly Ash Composites ...................................... 3
Flyash ........................................................... 43
Focused Ion Beam (FIB) .............................. 67
Forming ...................................................... 243
Fracture Toughness .................................... 185

## G
Geopolymer ................................................... 43

## H
High-energy X-ray Diffraction ................... 127
Hybrid Metal Matrix Composites ................ 91

## K
Kiss Bonds .................................................. 213

## L
Laser Deposition .......................................... 67

## M
Martensitic Transformation ........................ 101
Mechanical ................................................... 23
Mechanical Properties ................ 101, 157, 243
Mechanochemical Process (MCP) ............... 55
Metal Matrix Composites ........................... 115
Metal Nanocomposite ................................ 127
Metal-matrix Composites ............................ 81
Micro Cantilevers Testing .......................... 173
Micro-hardness ........................................... 101
Micro-silica Preform .................................... 81
Microhardness ............................................ 115
Microstructure ............................................ 101
MMCs ......................................................... 185
Multiscale Modeling .................................. 147

## N
Nanoindentation ......................................... 173
Nickel-Titanium-Carbon Composites .......... 67
Nondestructive Evaluation ......................... 137
Nondestructive Testing .............................. 213

## P
Percussion .................................................. 213
Phase Transformation .................................. 67
Polymer Blends .......................................... 157
Polymer Composites .................................. 147
Polymeric Fiber Composites ...................... 231
Processing Maps ............................................ 3

Properties ...... 23

# R
Redmud ...... 43

# S
Shape Memory Alloys ...... 101
SiC ...... 173
SiC Fibers Reinforced SiC Matrix ...... 173
Squeeze Infiltration ...... 81, 91
Steel/polymer/steel Sandwich ...... 243

# T
TaC-TaB$_2$ Composite ...... 55
Tantalum Carbide ...... 55
Thermal Expansion ...... 127
Thermosets ...... 157
Ti-Al Alloys ...... 255
Titanium Carbide ...... 255

# W
Wetting ...... 255

# Z
Zinc-Aluminum ...... 115